Solar Farms

The market and policy impetus to install increasingly utility-scale solar systems, or solar farms (sometimes known as solar parks or ranches), has seen products and applications develop ahead of the collective industry knowledge and experience. Recently however, the market has matured and investment opportunities for utility-scale solar farms or parks as part of renewable energy policies have made the sector more attractive. This book brings together the latest technical, practical and financial information available to provide an essential guide to solar farms, from design and planning to installation and maintenance.

The book builds on the challenges and lessons learned from existing solar farms, that have been developed across the world, including in Europe, the USA, Australia, China and India. Topics covered include system design, system layout, international installation standards, operation and maintenance, grid penetration, planning applications, and skills required for installation, operation and maintenance. Highly illustrated in full colour, the book provides an essential practical guide for all industry professionals involved in or contemplating utility-scale, grid-connected solar systems.

Susan Neill is Director of Training and Engineering at Global Sustainable Energy Solutions (GSES), Australia.

Geoff Stapleton is Managing Director of GSES and a part-time lecturer at the School of Photovoltaic and Renewable Energy Engineering, University of New South Wales, Australia.

Christopher Martell is Director of Operations and Engineering at GSES, Australia.

Earthscan Expert Series
Series editor: Frank Jackson

Solar:

Grid-Connected Solar Electric Systems
Geoff Stapleton and Susan Neill

Pico-solar Electric Systems
John Keane

Solar Cooling
Paul Kohlenbach and Uli Jakob

Solar Domestic Water Heating
Chris Laughton

Solar Technology
David Thorpe

Stand-alone Solar Electric Systems
Mark Hankins

Solar Farms
Susan Neill, Geoff Stapleton and Christopher Martell

Home Refurbishment:

Sustainable Home Refurbishment
David Thorpe

Wood Heating:

Wood Pellet Heating Systems
Dilwyn Jenkins

Renewable Power:

Renewable Energy Systems
Dilwyn Jenkins

Energy Management:

Energy Management in Buildings
David Thorpe

Energy Management in Industry
David Thorpe

Bioenergy:

Anaerobic Digestion
Tim Pullen

Solar Farms

The Earthscan Expert Guide to Design and Construction of Utility-scale Photovoltaic Systems

Susan Neill, Geoff Stapleton and Christopher Martell

LONDON AND NEW YORK

First published 2017
by Routledge
2 Park Square, Milton Park, Abingdon, Oxon OX14 4RN

and by Routledge
605 Third Avenue, New York, NY 10017

First issued in paperback 2021

Routledge is an imprint of the Taylor & Francis Group, an informa business

British Library Cataloguing-in-Publication Data
A catalogue record for this book is available from the British Library

Library of Congress Cataloging in Publication Data
Names: Neill, Susan, 1950- author. | Stapleton, Geoff, author. | Martell, Christopher, author.
Title: Solar farms : the Earthscan expert guide to design and construction of utility-scale photovoltaic systems / Susan Neill, Geoff Stapleton and Christopher Martell.
Description: London ; New York : Routledge, is an imprint of the Taylor & Francis Group, an Informa Business 2017. | Series: Earthscan expert series | Includes bibliographical references and index.
Identifiers: LCCN 2016025481| ISBN 9781138121355 (hbk) | ISBN 9781315651002 (ebk)
Subjects: LCSH: Photovoltaic power systems--Design and construction. | Building-integrated photovoltaic systems--Design and construction.
Classification: LCC TK1087 .N45 2017 | DDC 621.31/244--dc23
LC record available at https://lccn.loc.gov/2016025481

ISBN 13: 978-0-367-78399-0 (pbk)
ISBN 13: 978-1-138-12135-5 (hbk)

Typeset in Sabon
by Saxon Graphics Ltd, Derby

Contents

Illustrations

Figures

Boxes

Tables

Acronyms and abbreviations

AC	alternating current
alt_{EQ}	solar altitude during the equinox
alt_S	solar altitude during the solstice
ANSI	American National Standards Institute
ARENA	Australian Renewable Energy Agency
AS	Australian standard
a-Si	Amorphous silicon
AVR	Automatic voltage regulation
AWG	American wire gauge
BLM	Bureau of Land Management
BMS	Building management system
BoS	balance of system
BS	British Standards
CAD	computer-aided design
CB	circuit breaker
CCC	current-carrying capacity
CCTV	closed-circuit television
CdTe	cadmium telluride
CEC	Commission of the European Communities
CEFC	clean energy finance corporation
CfD	contract for difference
CIGS	copper–indium–gallium–diselenide
CSA	cross-sectional area
DC	direct current
DIN	German standards
DNI	direct normal irradiance
DNO	distribution network operator
DRED	demand response-enabled devices
ECT	equivalent cell temperature
EFI	earth fault interrupter
EIA	environmental impact analysis/assessment
ELV	extra low voltage
EMI	electromagnetic interference
EMS	energy management system
EN	European standards
EPBC	environmental protection and biodiversity conservation
EPC	engineering, procurement and construction
FCAS	frequency control ancillary services
FiT	feed-in tariff
GFDI	ground fault detection interrupter
GHGs	greenhouse gases

GHI	global horizontal irradiation
GIS	geographic information system
GSC	solar constant
GSES	global sustainable energy solutions
GUI	graphical user interface
GW	gigawatt
HMI	human–machine interface
HS	health and safety
HV	high voltage
I_{trip}	rated trip current
IEC	International Electrotechnical Commission
IEEE	Institute for Electrical and Electronics Engineers
Imp	current at maximum power point
IP	ingress protection
IREC	Interstate Renewable Energy Council
IRR	internal rate of return
Isc	short-circuit current
ISO	International Organization for Standardization
IT	information technology
JIS	Japanese Standards
LCF	local control facility
LCOE	levelised cost of electricity
LV	low voltage
MC	multi-contact
MCB	miniature circuit breaker
MET	meteorological
MJ	Mega-joule
MPP	maximum power point
MPPT	maximum power point tracker
MTBF	mean time between failures
MV	medium voltage
MVA	megavolt ampere
MW	megawatt
MWp	megawatt peak
NA	not applicable
NEC	National Electric Code
NEMA	National Electrical Manufacturers Association
NFPA	National Fire Protection Association
NOCT	nominal operating cell temperature
NPV	net present value
NREL	National Renewable Energy Laboratory
NSW	New South Wales
NZS	New Zealand standard
O&M	operations and maintenance
OLE	object linking and embedding
OPC	open platform communications
OSHA	occupational safety and health administration
PF	power factor

PLC	programmable logic controller
Pmp	maximum power point
POI	point of interface
PPA	power purchase agreement
PPP	public–private partnership
PR	performance ratio
PSH	peak sun hours
PV	photovoltaic
RAPID	regulatory and permitting information desktop
RCD	residual current device
RECs	renewable energy certificates
REIPP	renewable energy independent power producer procurement
RFP	request for proposals
ROW	right of way
RPS	renewable portfolio standard
RTU	remote telemetry unit
SCADA	supervisory control and data acquisition
SRECs	solar renewable energy credits
SRMC	short-run marginal cost
STC	standard test conditions
SWMS	safe work methods statements
TNO	transmission network operator
UK	United Kingdom
UL	Underwriters' Laboratory
UPS	uninterruptible power supply
USA	United States of America
UV	ultra-violet
VAR	volt–ampere reactive
Vmp	voltage at maximum power point
Voc	open-circuit voltage
WA	Western Australia
Wh	watt hour
Wp	watt-peak

Notes on the authors

Susan Neill has been involved in the Australian and international renewable energy industry for 30 years. Susan has extensive experience arising from her experience with the rapid market development in grid-connected solar since 2005. Susan had worked with international product manufacturers and distributors as well as providing EPC services before joining GSES, where she has held the position of training and engineering director since 2009.

Geoff Stapleton has been part of the renewable energy industry for over 30 years and has been instrumental in the development of industry training and capacity building in Australia and many other countries: Ghana, Sri Lanka, Malaysia, China and the Pacific region. Geoff's vast engineering experience has been instrumental in the development of Australian standards and guidelines for the renewable energy industry. Geoff has been a part-time lecturer at the University of New South Wales, Australia for the past 15 years.

Christopher Martell has held the position of Principal Engineer for GSES since 2014. Christopher has extensive experience through his multidiscipline engineering qualifications, including the design and implementation of large-scale grid-connected solar systems. Christopher has contributed to the vast information base required to write this publication and to ensure currency for the equipment and practices included.

Preface and acknowledgements

The world of renewable energy has expanded from the time when it was considered only as a potential source of energy for niche applications: renewable energy is now touted as one of the foundations for mitigating global climate change. Solar photovoltaics are able to be deployed to meet the insatiable power demands ranging from a solar lantern through to powering populations in cities.

The impetus of solving the world's global warming problem and the need to find suitable alternative energy sources has contributed to and highlighted the enormous growth in the photovoltaic industry since 2009. Since that time, the prices for solar modules have reduced by a factor of five and the price for solar systems has reduced by a factor of three. The 2014 IEA Roadmap states that the cumulative installed capacity of solar PV has grown at an average rate of 49% per year in the years up to 2014.

A stark indication of this industry's stellar growth can be seen when comparing the total installed capacity of solar PV at the end of 2009 at 23 GW compared to the figure of 227 GW installed capacity at the end of 2015.

The solar farm, or utility-scale solar installation, is now an established part of international power supply landscape as well as being the catalyst for the introduction and adaptation of financial instruments and funding models so that the services for solar farms are part of mainstream banking, finance and the law.

This publication describes the solar technology used in solar farms; the technology's performance characteristics; all enabling equipment used in a solar farm; the extensive system design required; the supporting technical, social and environmental aspects; how to estimate a solar farm's performance and financial metrics.

Susan Neill, Geoffrey Stapleton and Christopher Martell from Global Sustainable Energy Solutions Pty Ltd (GSES) have extensive knowledge and experience in grid-connected solar systems and these systems' engineering, design, compliance and performance. As the authors of this publication, they have welcomed the opportunity to produce this *Solar Farms* publication, as this industry continues its international growth.

Given the rapid growth in the international market for solar farms, extensive time and resources were necessary to research and document the status of this constantly evolving market. Kayla Inglis in GSES's Sydney office has worked extensively on this publication: collecting the technology and market information, researching, writing content and developing the chapters to meet the broad, international context of this topic. GSES thanks those industry participants who have supported the development of this publication and provided information and images and we trust these are correctly acknowledged.

1

Introduction

Background

Context

In modern society, electricity is considered a fundamental public commodity as it is used in almost every aspect of life (see Figure 1.1). Electricity is essential for human well-being, and plays a key role in promoting social development and economic growth. However, growing electricity demand in the modern and developing world, coupled with the looming danger of climate change, means that it is becoming more and more imperative for cleaner renewable energy technologies to replace traditional fossil-fuelled generators.

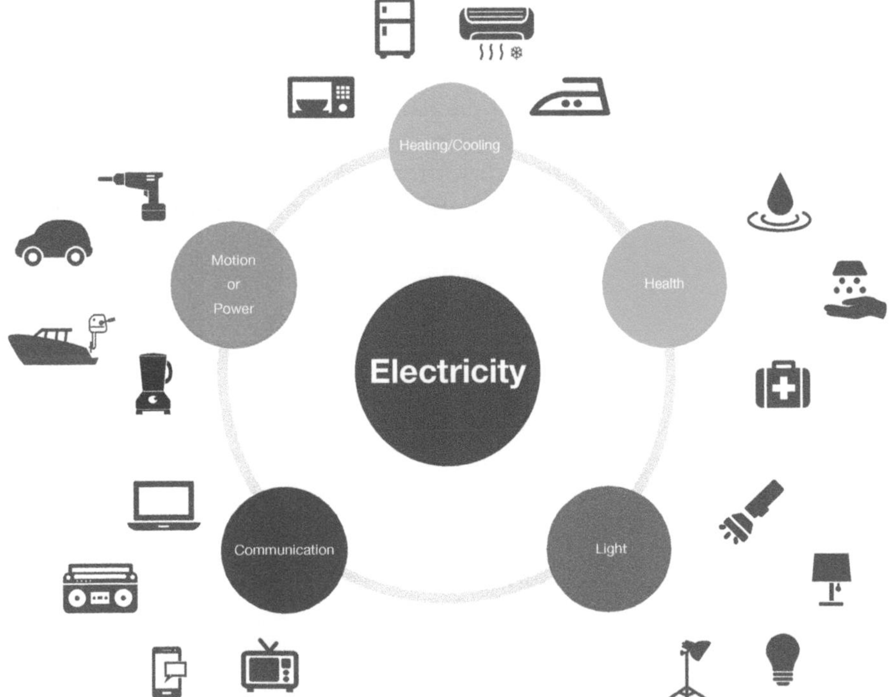

Figure 1.1 Uses of electricity in our daily lives.

Source: Global Sustainable Energy Solutions

In response to the threat of climate change, many countries around the world have committed to reduce greenhouse gas (GHG) emissions through the implementation of renewable energy targets, emissions reduction plans, carbon pricing, etc. This commitment, coupled with reduction in the price of photovoltaic modules, is driving investment in solar photovoltaic (PV) technologies. There are currently 164 countries (up from 43 countries in 2005) with a stated renewable energy target; the majority of these are countries with developing and emerging economies (IRENA 2015). Analysis has shown that in order to keep the global temperature rise within safe limits, renewable energy must account for 36% or more of global electricity generation by 2030, requiring USD 550 billion in annual investment (IRENA 2014). As a result of the 2015 Conference of Parties (COP21) in Paris, 195 countries have agreed to keep warming below 1.5–2°C above pre-industrial levels to reduce the risks and impacts of climate change. One hundred and eighty six of these countries have pledged to reduce GHG emissions through aggressive renewable energy targets, and will be reviewed and monitored in five-year cycles.

Demand

Electricity demand has increased steadily since the Industrial Revolution and is expected to continue growing over coming decades (see Figure 1.2). While countries such as those in the European Union, the United States and Japan are expected to reduce overall demand by 2040, other countries, particularly China and India, are expected to grow quite significantly.

This continuing growth in demand calls for aggressive action in order to keep atmospheric CO_2 within internationally agreed levels. With pledges from almost every country around the world committing to take real action on climate change, Figure 1.3 shows that renewable energy technologies are expected to make up the majority of the generation mix worldwide by 2040.

Although electricity demand is on the rise, there are still approximately 1.1 billion people without access to electricity, most of whom are concentrated in Africa and Asia. In addition to that, there are still 2.9 billion people (840 million people in India and about 450 million people in China) that rely on wood or

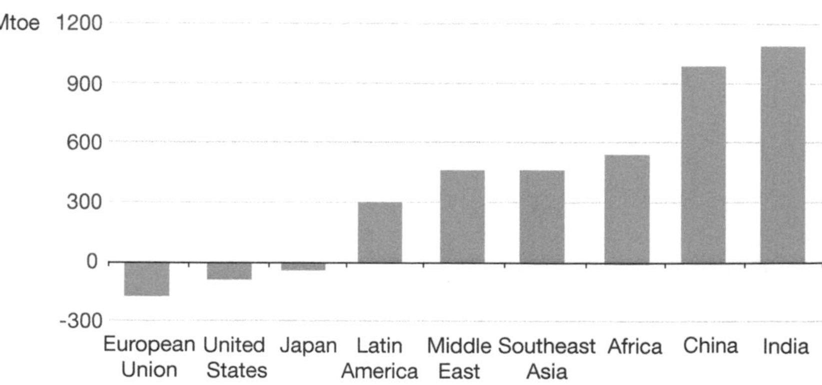

Figure 1.2 Expected change in energy demand in selected regions (2014–40).

Source: OECD/IEA

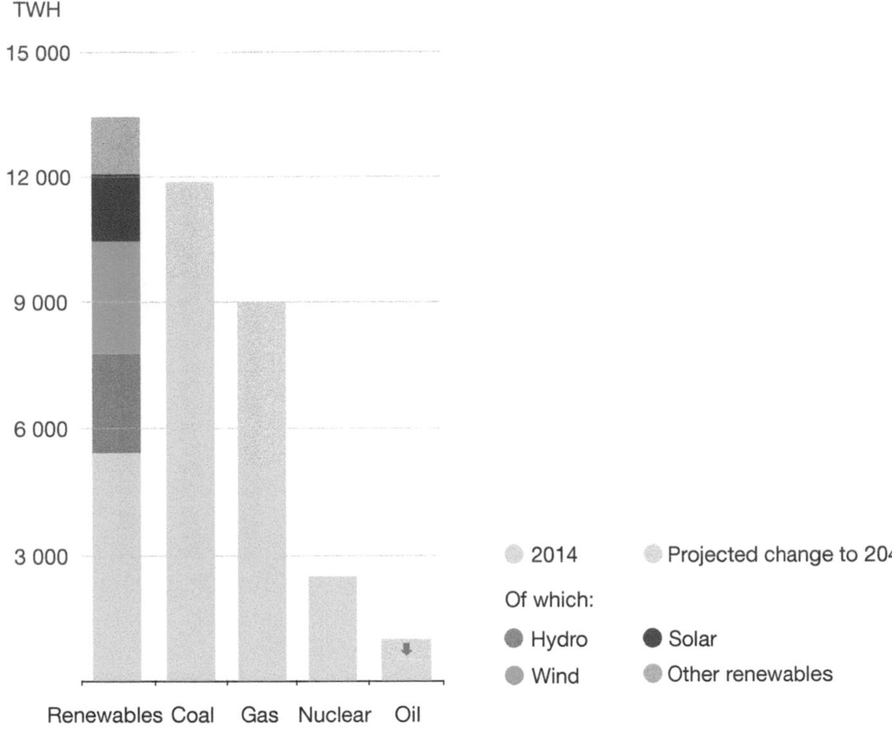

Figure 1.3 Current and projected global electricity generation by source.

Source: OECD/IEA

other types of biomass for cooking and heating, which is causing about 4.3 million premature deaths from indoor air pollution each year (The World Bank Group 2015). Clean renewable energy technologies like solar PV can play a key role in improving electricity access to remote communities in these areas. The International Energy Agency predicts that the number of people without access to electricity will decline to around 810 million in 2030 and 550 million in 2040 which will be about 6% of the global population at that time (IEA 2015). As more and more parts of the world have access to electricity, the utility-scale solar market is likely to follow and therefore utility-scale solar is expected to make up a larger portion of the world's electricity generation mix.

PV applications

Solar PV technology enables electricity to be generated directly from sunlight. Electric power can be generated using large solar farms (centralised grid-connected) or by solar installations on buildings (decentralised grid-connected), as well as at locations not connected to utility grids (off-grid applications). A breakdown of the global market share for each of these applications is given in Figure 1.4.

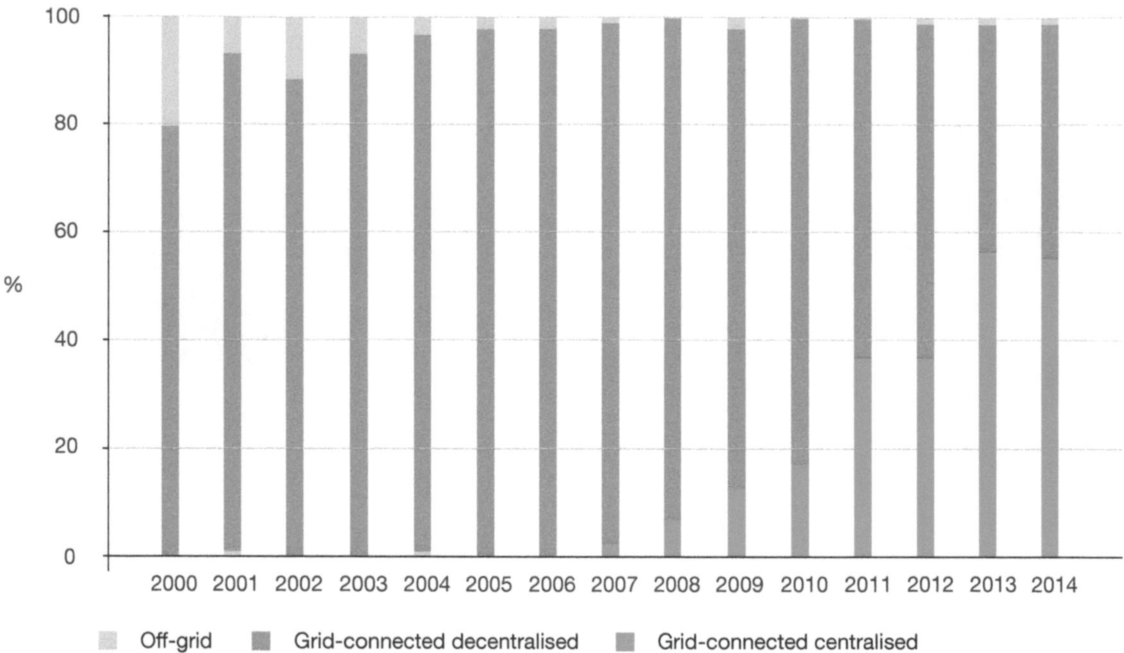

Figure 1.4 Global market share of off-grid, decentralised grid-connected and centralised grid-connected PV systems.

Source: IEA PVPS

Centralised grid-connected systems

Centralised grid-connected systems are large-scale PV systems, also known as solar farms. The definition of a solar farm varies, based on its scale, the mounting structure, type of grid connection, etc., and may be known as a solar park, solar plant, utility-scale solar system, etc. depending on where you are. These systems are typically ground mounted and are built to supply bulk power to the electricity grid like any other centralised power station. For this publication, a solar farm is defined as a ground-mounted, centralised, grid-connected PV system of the scale 10 MWp or above. Looking at Figure 1.4, utility-scale systems made up a very insignificant portion of global market share of PV systems since up until 2007, and has rapidly grown since; now making up a larger share than decentralised PV systems. Declining costs of PV technology, coupled with government policies targeting large scale renewable energy, have allowed utility-scale solar to become more and more competitive with other forms of decentralised electricity generation, driving rapid deployment in many countries across the globe.

Due to the large scale of these solar systems, the processes involved in developing these projects are far more complex and rigorous than that for smaller, decentralised PV systems. Successful projects require high capital investments, expertise in many areas, thorough planning, careful design, consultation with many stakeholders and ongoing maintenance.

Figure 1.5 105.56MW Perovo Solar Park (Crimea, Russia).

Source: Activ Solar

With the help of policy support in the form of renewable energy targets, deployment of solar farms worldwide is increasing at a faster rate than information and training specific to these larger-scale systems is being made available. This book aims to provide relevant and thorough information on the planning, design, construction, commissioning, operation and maintenance of utility-scale solar systems.

Decentralised grid-connected PV systems

Decentralised grid-connected PV systems may be installed to meet a residential, commercial or industrial requirement (typically on rooftops). These systems generate electricity that is either consumed in the building or sold onto the grid – with price determined by policymakers or utilities. These systems can be installed as a cost-effective option to users who consume energy during the day, with the excess generation able to be sold to the grid. Decentralised grid-connected PV systems made up the bulk of the global markets share of PV systems until 2013, when centralised systems overtook them.

Off-grid PV systems

Off-grid PV systems can be used for domestic applications to provide electricity (lighting, refrigeration, etc.) to households that are not connected to the utility grid; or for non-domestic applications to provide electricity for commercial application in remote areas for things like telecommunications, water pumping, vaccine refrigeration and navigation. Off-grid PV systems are usually installed in remote areas where connection to the grid is either not possible, or very expensive. While the market share of off-grid systems is tiny compared to grid-connected systems (see Figure 1.4), this market is continuing to grow internationally, e.g. in China, Australia, India and other countries in which it is too difficult and expensive to extend the grid.

Figure 1.6 Residential grid-connected PV system.

Figure 1.7 Installation of an off-grid PV system in Ghana.

Source: Global Sustainable Energy Solutions

Why a solar farm?

If policymakers are serious about climate change action, renewable energy policy must incentivise the procurement of utility-scale renewable energy. Distributed generation can play an important role in decarbonising the electricity sector, reduce transmission losses, and reduce the need for transmission and distribution infrastructure to be located close to electricity demand centres. However, distributed generation is limited in terms of scale and to suitable rooftops. Utility-scale solar farms take advantage of economies of scale by deploying multi-megawatts of solar modules and inverters in large open spaces.

Compared to other types of centralised power stations, solar farms cause minimal disturbance to the land:

- They can be fully decommissioned at the end of their lifetime, allowing the land to be restored to its original use.
- Once the solar farm has been constructed, there is little ongoing disturbance from humans or machinery on the site; the land beneath the PV modules can be used for limited agriculture.
- Solar farms are often built on land of limited financial or community value, such as previously developed land, contaminated or industrial land – so-called 'brownfield' sites.

Solar farms can also have a range of social and economic benefits, including:

- diversification of the electricity generation mix – improving energy security;
- stimulation of the local economy by providing jobs and training in the local area, and supporting local suppliers and dealers; and
- reduction of energy import dependency.

With solar technology costs decreasing so rapidly, solar farms are becoming increasingly competitive with other centralised generators. However, there is still a way to go before they are competitive without policy support. Depending on the political environment, solar farms have the potential to play a very promising role in helping to meet national and international renewable energy and emissions reduction targets.

Solar farms in the global market

While the global market for solar farms is changing year by year, a summary of this industry at the time of writing is included in this chapter. This historical information emphasises the rapid market growth over a few years, pinpoints those countries that are major players in this market, and emphasises the relationship between a country's energy policy and the growth of this market.

As noted earlier, the global PV market share of solar farms (centralised PV systems) has increased to comprise 55% of the PV capacity installed during 2014. Having centralised PV systems exceeding the installed capacity of decentralised systems is not the case for all countries. Figure 1.9 shows that this trend is evident across the Americas, the Middle East and Africa region, as well as in the Asia Pacific region, while Europe remains predominantly rooftop solar.

The history of the extraordinary market growth of solar farms (centralised solar PV systems) is summarised below.

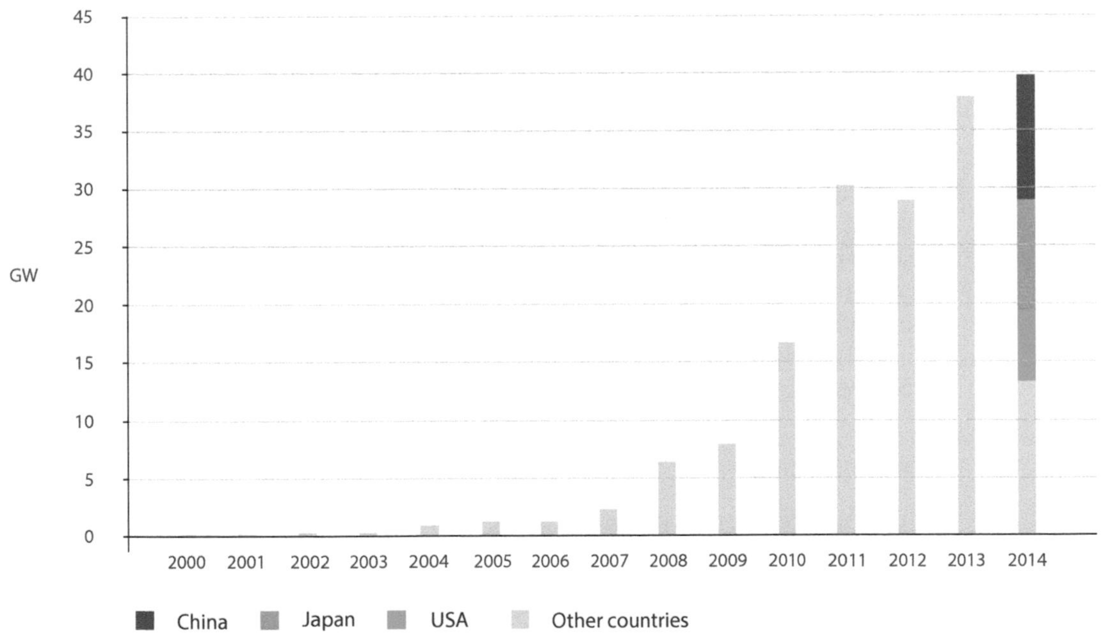

Figure 1.8 Evolution of annual PV installations (GW).

Source: IEA PVPS

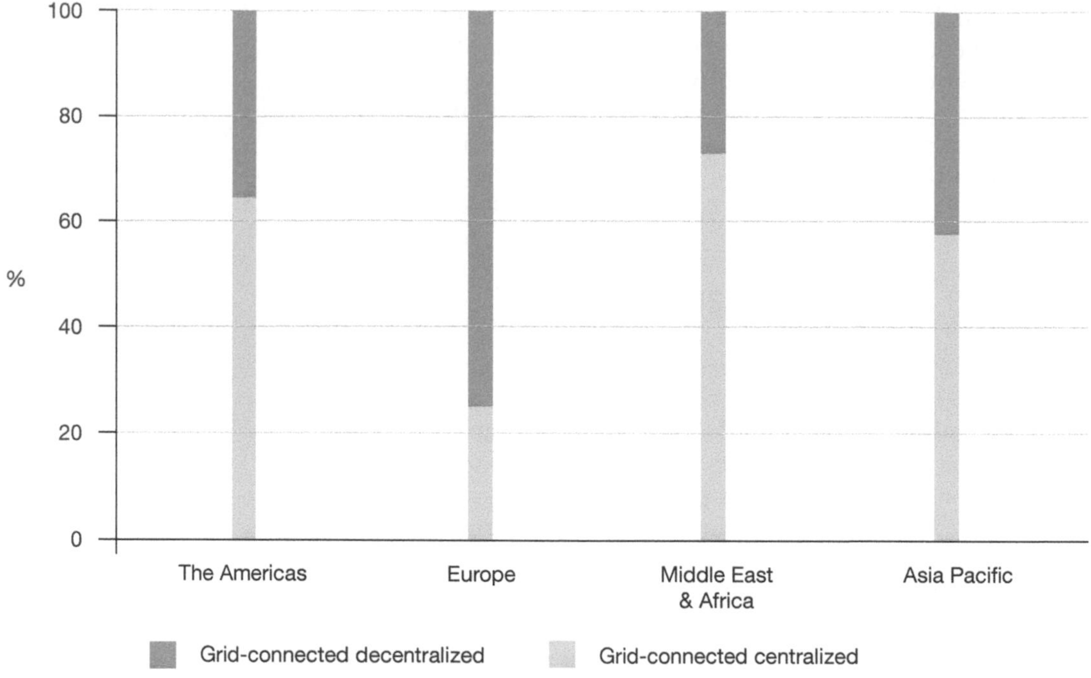

Figure 1.9 Grid-connected centralised and decentralised PV installations by region in IEA PVPS countries.

Source: IEA PVPS

The Americas

The PV market in the Americas is largely dominated by the US, followed by Canada and Chile, installing a combined total of 7.4 GW in 2014, with a total installed capacity of 21 GW. While the US PV market has historically been majority distributed systems, by the end of 2014 utility-scale systems made up 63% of the total installed capacity, with some of the largest solar farms in the world coming online during 2015. Looking at Table 1.1, the US is home to seven of the ten largest solar farms in the world. This has been largely driven by requirements under the Renewable Portfolio Standard (RPS) and from declining installation costs, increasing competitiveness with other generators like natural gas. By early 2015, more than 14 GW of projects were scheduled to come online before changes to federal tax credit at the end of 2016.

Asia Pacific

The Asia Pacific region dominated the PV market in 2013–14, installing 24 GW in 2014, totalling 63.5 GW of installed capacity by the end of 2014, most of which was made up by China, followed by Japan and Australia. While the majority of PV installations across Japan and Australia were rooftop systems (5 GW compared to 3 GW in Japan, and 803 MW compared to 83 MW in Australia) thanks to government FiT policies, utility-scale systems made up the vast majority of installations across China (8,550 MW compared to 2,100 MW) during 2014

Table 1.1 Top ten largest solar farms in the world

1. Solar Star (I and II)	Capacity: 579 MW_{AC} Location: California, USA Year: 2015 Module type: thin film Mounting structure: single-axis tracking
2. Topaz Solar Farm	Capacity: 550 MW_{AC} Location: California, USA Year: 2015 Module type: thin film Mounting structure: fixed ground mounted
3. Desert Sunlight Solar Farm	Capacity: 550 MW_{AC} Location: California, USA Year: 2015 Module type: thin film Mounting structure: fixed ground mounted
4. Copper Mountain Solar Facility	Capacity: 458 MW_{AC} Location: Nevada, USA Year: phase 3 completed in 2015, final phase (additional 94 MW_{AC}) scheduled for completion by end of 2016 Module type: thin film Mounting structure: fixed ground mounted
5. Longyangxia Dam Solar Park	Capacity: 320 MW_{AC} Location: Qinghai, China Year: 2013 Module type: polycrystalline Mounting structure: fixed ground mounted
6. Cestas Solar Farm	Capacity: 300 MW_{AC} Location: Bordeaux, France Year: 2015 Module type: polycrystalline Mounting structure: single-axis tracking
7. Agua Caliente Solar Project	Capacity: 290 MW_{AC} Location: Arizona, USA Year: 2014 Module type: thin film Mounting structure: fixed ground mounted
8. California Valley Solar Ranch	Capacity: 250 MW_{AC} Location: California, USA Year: 2013 Module type: monocrystalline Mounting structure: single-axis tracking

Table 1.1 *Continued*

9. Charanka Solar Park Facility	Capacity: 224 MW_{AC} Location: India Year: 2013 Module type: thin film Mounting structure: fixed ground mounted
10. Mount Signal Solar Project	Capacity: 206 MW_{AC} Location: California, USA Year: 2014 Module type: thin film Mounting structure: single-axis tracking

(IEA PVPS 2015). However, while the majority of utility-scale capacity have been built in the sunny western regions of China, the transmission infrastructure has not been upgraded to transmit all of this generation to the densely populated areas in the south and east, leaving many of these solar farms underutilised.

Europe

European countries led PV development until 2012, after which PV installations across Europe declined, while installations in the rest of the world have been grown rapidly. During 2014, European countries installed a combined 7 GW of PV, bringing it to a cumulative capacity of 89 GW, which remains the highest of any other region. In 2014 the UK overtook Germany in annual installations, followed by France. Germany installed over 7 GW of PV for three years in a row until the end of 2012 thanks to a combination of long-term stability of support schemes; confidence of investors; and keen residential, commercial and industrial building owners. The majority of support schemes across Europe have been tailored towards distributed systems (the UK was the only European country to install a significant amount of utility-scale capacity during 2014). Hence rooftop systems have dominated the PV market in Europe to date.

Middle East and Africa

Although the Middle East and Africa have an excellent solar resource, this part of the world is behind other regions in terms of PV uptake. However, since 2014 things have been changing, with South Africa installing 800 MW, the majority of which is utility-scale driven by the Renewable Energy Independent Power Producer Procurement (REIPP) programme. The first utility-scale system was connected to the grid in Israel at the end of 2014, making up 37.5 MW of the total 200 MW of PV systems installed that year. The majority of capacity installed in Israel during the following two years is expected to come from large centralised PV systems. While PV has not taken off in these regions yet, many utility-scale systems are planned across various African countries to replace or complement existing diesel generators, while many countries in the Middle East have launched tenders to procure large-scale solar over the next few years.

Business models

Any solar farm project requires a strong and detailed business plan. This business plan must include detailed and specific documentation to ensure that the financing for this project, either through securing the required loan funds or equity financing can be substantiated.

The supporting information required by potential lenders will include the financial modelling for the proposed solar farm. The business credentials and success of a solar farm will therefore be the product of a close working relationship with an experienced solar farm developer and suitably experienced financial entities. It is imperative that the technical deliverables of the project are closely managed and are undertaken by an experienced EPC (engineering, procurement and construction) company or group; and the financing parties must fully understand the nature, risk and market deliverables of investing in a solar farm.

It is important that any business model for a solar farm includes the financial benefits arising from available tax credits, accelerated depreciation, etc. and any rebates or grants offered for solar farms. The overall financial assessment of the project will also include the income to the solar farm from the long-term sale of the power to the utility or network involved.

Any business model developed for a solar farm will seek to optimally combine the interests of the project's financiers and investors and the interests of those parties intending to use (i.e. sell or consume) the solar farm's output.

Most often, the project will be owned and operated by a company backed by investors: a utility rather than solar equipment suppliers or developers. This investor-owned company will be responsible for negotiating power purchase agreements with utilities, consumers or businesses to buy the power produced, and hiring contractors to carry out the engineering, procurement, construction and operations of the project. The company will earn a return based on the sale of electricity and from any government subsidies that may be available. However, it is unlikely that this company will be able to finance the whole project through equity and will almost always seek debt finance from banks, who earn a return through interest. More information on finance options is given in Chapter 3.

Bibliography

Cronshaw, Ian. 'World Energy Outlook 2015'. OECD/IEA, 2015.

IEA. 'Energy Access Projections'. 2015. www.worldenergyoutlook.org/resources/energydevelopment/energyaccessprojections/ (accessed 17 December 2015).

IEA PVPS. 'A Snapshot of Global PV Markets 2014'. 2015.

IEA PVPS. 'Trends 2015 in Photovoltaic Applications'. 2015.

IRENA. 'Renewable Energy Target Setting'. 2015.

IRENA. 'REthinking Energy'. 2014.

Nuttall, Nick. 'Historic Paris Agreement on Climate Change 195 Nations Set Path to Keep Temperature Rise Well below 2 Degrees Celsius'. 2015. http://newsroom.unfccc.int/unfccc-newsroom/finale-cop21 (accessed 16 December 2015).

REN21. 'Renewables 2015 Global Status Report'. 2015.

World Bank Group. Access to Electricity (% of population). 2015. http://data.worldbank.org/indicator/EG.ELC.ACCS.ZS/countries?display=map (accessed 17 December 2015).

World Bank Group. Energy Overview. 2015. www.worldbank.org/en/topic/energy/overview#1 (accessed 17 December 2015).

2

Photovoltaic technology

This chapter describes the principles behind solar technology in general, providing a basis for the understanding of the solar resource and how this resource is able to be converted to electricity. Further, this chapter includes a description of all of the physical components that make up a utility-scale photovoltaic (PV) system. This will provide a foundation for the understanding of later chapters that describe the planning, design, construction and operation of the system involving these components.

Solar radiation

Sunlight is a form of energy that is emitted from the sun, also known as solar radiation. It must travel through the earth's atmosphere before it can be absorbed by photovoltaic (PV) cells and converted into electrical energy. The amount of electrical energy generated from a PV cell is directly proportional to the input solar radiation. The amount of solar radiation that reaches a PV cell varies depending on the location (latitude), time of day and time of year, as well as the tilt and orientation of the PV module with respect to the position of the sun in the sky. When planning and designing a solar farm, it is critical to use this information to select an appropriate site, and to design the system so that the modules are oriented in such a way as to maximise the output of the solar farm.

The solar resource

The solar radiation at the top of the earth's atmosphere, known as the solar constant (G_{SC}), is approximately 1.367 kW/m^2. However, as the solar radiation passes through the earth's atmosphere it is reduced to a peak value of approximately 1 kW/m^2. The amount of radiation that actually reaches the earth's surface depends on:

- Gases in the air absorbing different wavelengths, affecting the range of wavelengths of light that reaches the surface and can be absorbed by the PV cell. PV cells typically absorb light in the visible light spectrum.

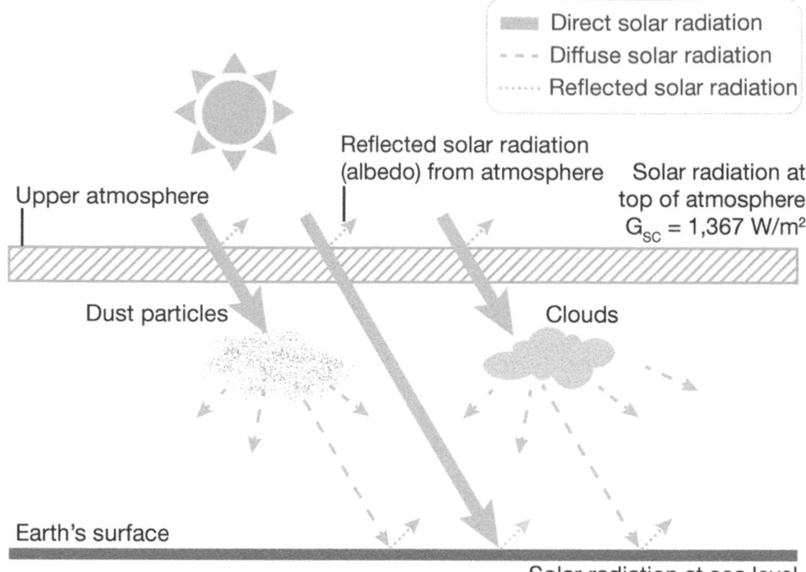

Figure 2.1 Albedo, direct radiation and diffuse radiation.

Source: Global Sustainable Energy Solutions

- Reflection and scattering of light (*albedo*), affecting the amount of *direct* (unscattered) and *diffuse* (scattered and less intense) light that reaches the surface (see Figure 2.1). PV cells absorb both direct and diffuse radiation, however since direct radiation is more intense, the output of a PV module will be higher when there is less scattering of light, and more direct sunlight reaching the surface of the PV cell.
- The amount of atmosphere that the solar radiation travels through (*air mass*). The air mass is directly proportional to the angle of the sun in the sky, therefore air mass is maximised in the morning and afternoon, and is at a minimum during the middle of the day when the sun is directly overhead. The greater the air mass, the less solar radiation reaches the earth's surface because the atmosphere absorbs, reflects and scatters solar radiation.

Measuring solar radiation

Solar radiation is typically measured by the amount of radiation that falls in an instant, over a day or over a year, or as the timeframe in which 1 kW of radiation will land on an area of 1 m^2. Site-specific solar radiation is required to predict the solar farm's performance in situ. There will be various sources of long-term solar radiation data available for most locations around the world. However, the reference points for this quoted radiation must be correlated to the intended solar farm installation location. Therefore accurate measurements and/or calculations of site-specific solar radiation are needed in order to design and size system components, and in order to make predictions of the system yield.

Solar radiation data is also one parameter making up the system performance data collected once the solar farm is in operation. This information is required to evaluate and predict the system's performance.

It is important to understand and distinguish between the different terms and units used to express solar radiation.

- Irradiance is a measure of solar power per unit area at any moment in time. It is a measurement of instantaneous power and is measured in W/m^2 (or kW/m^2).
- Irradiation is the total amount of solar energy per unit area received over a given period (e.g. a day, a month or a year). It is a measurement of power over time and is measured in MJ/m^2 or Wh/m^2 (or kWh/m^2). Note that there are 3.6 MJ in 1 kWh.
- Peak sun hours (PSH) is the number of hours for which a site will receive solar irradiance at the peak value 1 kW/m^2. The number of PSH for a site is equivalent to the total irradiation received at the site during the day, and is illustrated in Figure 2.2.

Sun geometry

When designing a solar farm, in order to face the PV modules in a direction to maximise incident solar radiation, it is critical to consider the position of the sun throughout different times of day due to the earth's rotation and throughout the year due to the earth's orbit around the sun (Figure 2.3). For a solar farm, the modules should be positioned at the angle and direction that maximises the annual energy output.

The position of the sun relative to an observer can be defined by two angles:

1 Altitude angle is a measure of the height of the sun in the sky. It is the vertical angle between the sun and the horizontal plane (i.e. the horizon or the ground), and is always an angle between 0° and 90°, in the northern or southern direction (Figure 2.4). The altitude angle will vary throughout the day and throughout the year: it will be higher at noon than during the mornings and evenings, and higher in summer than in winter.

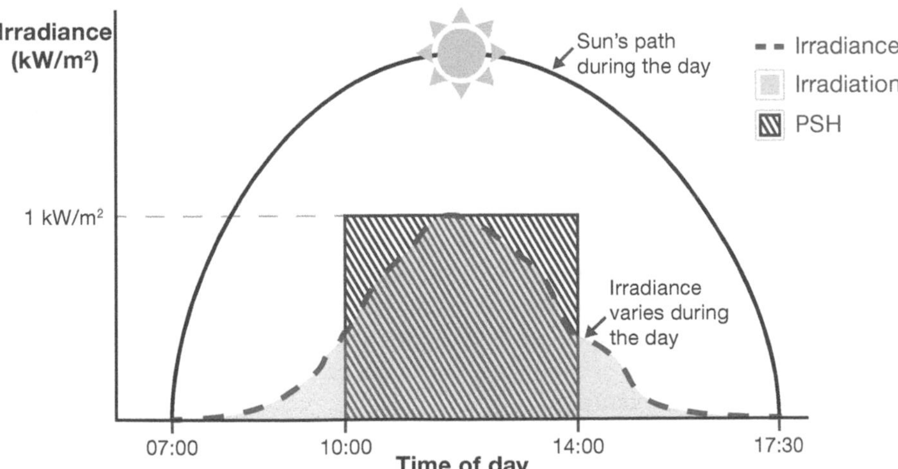

Figure 2.2 Schematic representation of PSH. On this particular day, this site is receiving 4 peak sun hours.

Source: Global Sustainable Energy Solutions

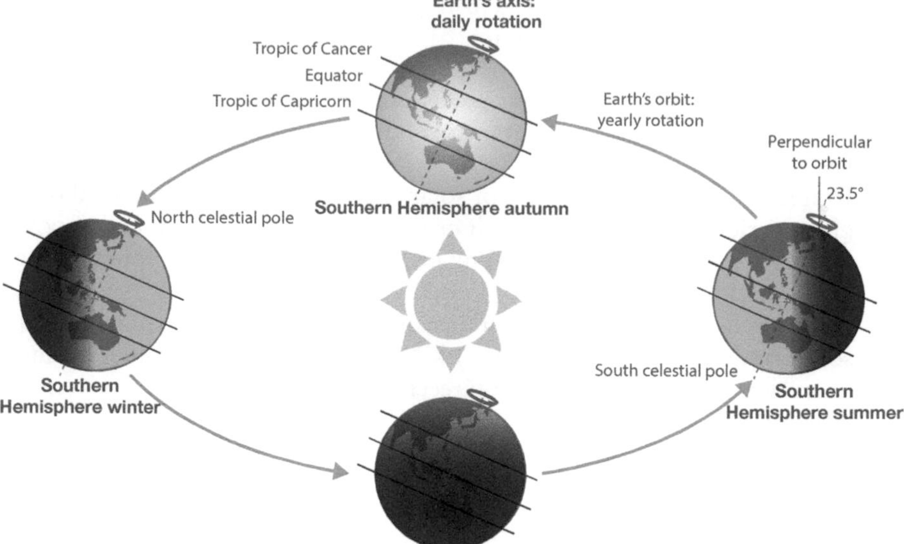

Figure 2.3 The orbit of the earth around the sun, combined with the earth's tilt, creates the seasons, causing the sun to move north or south in the sky.

Source: Global Sustainable Energy Solutions

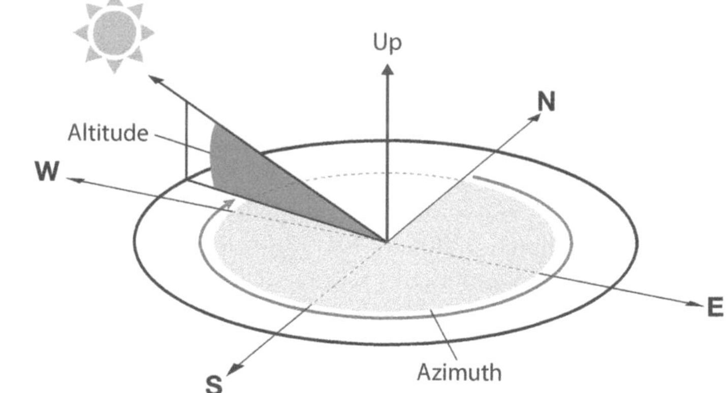

Figure 2.4 A visual representation of the sun's altitude and azimuth.

Source: Global Sustainable Energy Solutions

2 Azimuth angle measures the direction of the sun on a horizontal plane with respect to north (Figure 2.4). It is measured in a clockwise direction from true north (0°), and varies as the sun moves from east to west throughout the day.

Throughout the year the sun moves between the tropic of Cancer and the tropic of Capricorn, located at latitudes of 23.45° north and 23.45° south, respectively. A solstice occurs when the sun is above either of the tropics: whether it is a winter or summer solstice depends on which hemisphere the observer is in (Figure 2.5). An equinox occurs when the sun is half way between the tropics at the equator, located at latitude 0°.

The solar altitude (alt_{EQ}) during the equinox depends on the latitude of the observer and can be calculated using the following formula:

a. Southern Hemisphere

Summer (southern) solstice
Equinoxes
Winter (northern) solstice
N
S
E
W

b. Equator / Tropics

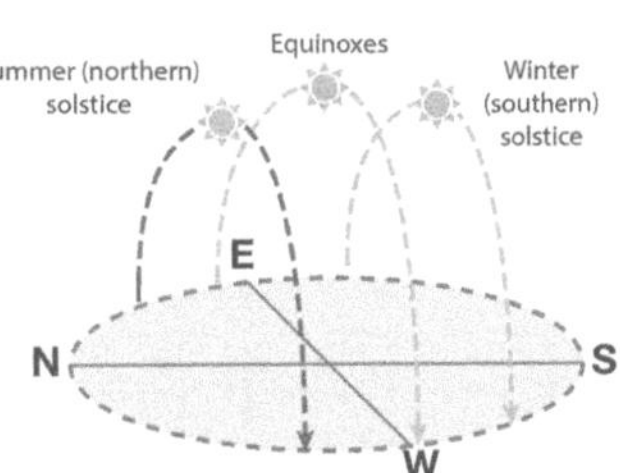

c. Northern Hemisphere

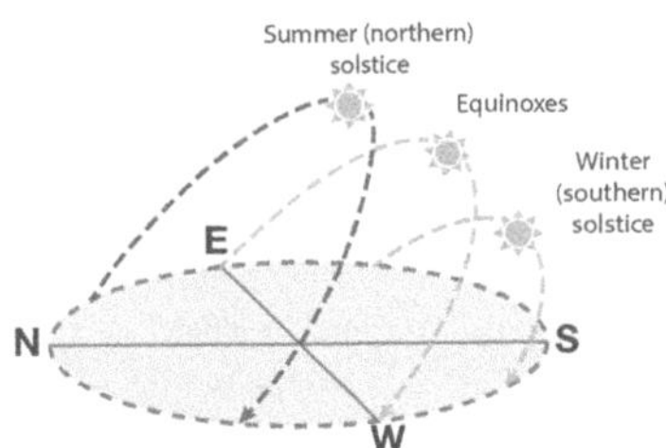

Figure 2.5 (a) In the southern hemisphere the sun tracks the highest path during the summer (southern) solstice and the lowest path during the winter (northern) solstice. (b) For tropical location located close to the equator, the sun remains high in the sky throughout the entire year, but is highest during the equinox. (c) In the northern hemisphere, the sun tracks the highest path during the summer (northern) solstice and the lowest path during the winter (southern) solstice.

Source: Global Sustainable Energy Solutions

$$alt_{EQ} = 90° - \text{latitude (in degrees)}$$

The solar altitude (alt_S) during the solstices also depends on the latitude of the observer and can be calculated using the following formula:

$$alt_s = 90° - \text{latitude (in degrees)} \pm 23.45°$$

Whether to add or subtract depends on whether it is the summer or winter solstice. For the summer solstice (the sun and the observer are in the same hemisphere), add 23.45° and, for the winter solstice (the sun and the observer are in different hemispheres), subtract 23.45°.

Once the position of the sun is known, the optimal tilt angle and orientation for a PV module can be calculated to maximise the amount of irradiation received. The maximum amount of radiation received at any given time occurs when the sun's rays are perpendicular to the PV modules. While a solar tracker will give the maximum output by ensuring that the modules move with the sun, given the high capital cost and maintenance requirements of tracking technology, it is often not cost-effective. If the modules are to be mounted in a fixed, stationary position then the modules should be tilted at an angle to maximise the average annual output. To achieve this, the modules should be tilted to face the equator during solar noon on the equinoxes. Further details on selecting the optimal tilt angle, and exceptions to this general rule, are given in Chapter 4.

Photovoltaic technologies

Photovoltaic cells

PV cells are made from semiconducting materials, which are mostly silicon-based. These materials are neither conductors nor insulators, but are able to conduct electricity under certain conditions. A silicon PV cell can conduct

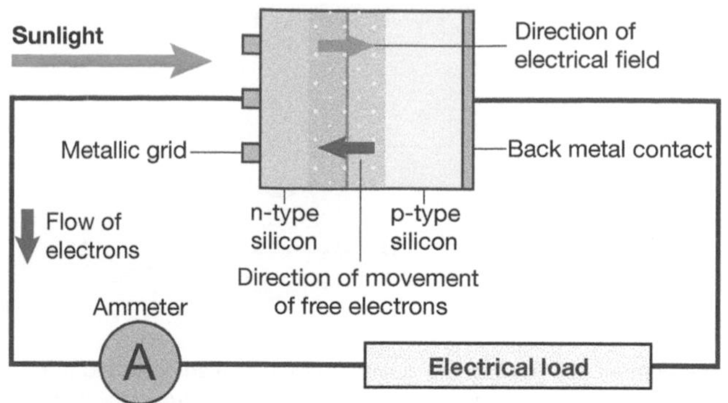

Figure 2.6 By combining the photoelectric effect and the electric field of the p–n junction, electrons can be induced to move inside the cell and create a current flowing in an external wire.

Source: Global Sustainable Energy Solutions

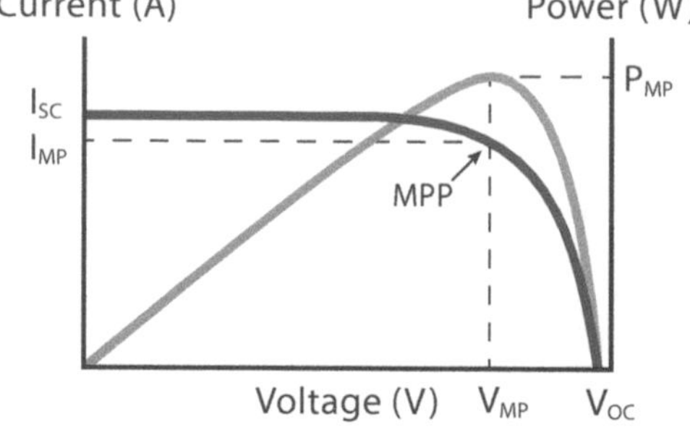

Figure 2.7 A characteristic I–V curve of a PV cell, with power output plotted for each point on the curve.

Source: Global Sustainable Energy Solutions

electricity when sunlight hits the surface, because impurities have been added to the silicon by a process called doping. In very basic terms, these impurities allow the electrons to move around once excited by an incident photon (particle of sunlight), creating current, allowing the cell to conduct electricity.

The I–V curve of a photovoltaic cell

In order to maximise the electrical output of a solar array, it should be designed so that it operates at maximum efficiency. The electrical output of a PV cell depends on where it is operating on the I–V curve, which shows the relationship between current and voltage of the electricity generated by the cell, illustrated in Figure 2.7.

The key parameters that define the core electrical characteristics of a PV cell are:

- Short-circuit current (I_{SC}): the current during a short circuit, i.e. when the resistance is zero. This is the maximum current output of the PV cell.
- Open-circuit voltage (V_{OC}): the voltage during an open circuit, i.e. when the resistance is at a maximum. This is the maximum voltage output of the PV cell.
- Maximum power point (MPP): the point on the I–V curve at which the PV cell will generate the most power.
- Maximum power (P_{MP}): the maximum power output of the PV cell.
- Voltage at PMP (V_{MP}): the PV cell voltage at the MPP of the I–V curve.
- Current at PMP (I_{MP}): the PV cell current at the MPP of the I–V curve.

It is important to operate the cells at their MPP, but this is not automatically achieved by the PV cells. Generally, the inverter connected to the PV array will contain a maximum power point tracker (MPPT) that ensures that the array operates at its MPP.

PV modules

PV modules are made up of a number of PV cells connected in series. The I–V characteristics of a PV module depend on the I–V characteristics of the individual PV cells that make up the module.

Connecting identical PV cells in series results in the same current output as a single cell but having a combined voltage output. This means that the I–V curve of the combined PV modules will have the same I_{SC} but an increased V_{OC} (Figure 2.8). The more cells connected in series, the higher the power output of the module will be.

As operating conditions have an effect on module performance, a standard set of operating conditions needs to be used to allow for a performance comparison between different modules. There are two main sets of operating conditions used for indicating the performance of a module: STC and NOCT.

Standard test conditions

Under international standards, all modules are tested at the following standard test conditions (STC):

- cell temperature of 25°C;
- irradiance of 1,000 W/m^2; and
- air mass of 1.5.

Nominal operating cell temperature

Many module manufacturers will provide data showing the performance of their PV modules at the nominal operating cell temperature (NOCT) as explained below. This shows the reduced performance of the module owing to the higher cell temperature and is a more accurate indication of actual cell performance (Figure 2.9).

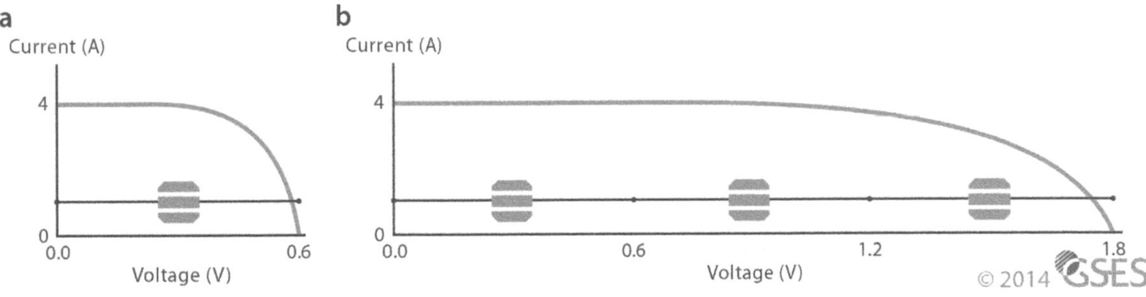

Figure 2.8 (a) The I–V characteristics of a single cell; (b) the I–V characteristics of three of those cells connected in series.

Source: Global Sustainable Energy Solutions

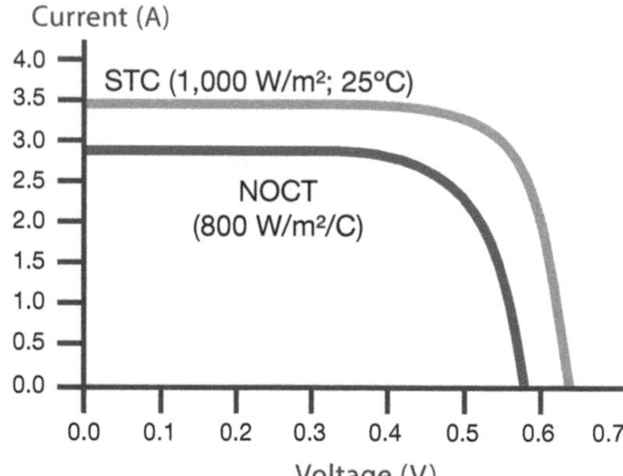

Figure 2.9 Typical PV cell performance curves under different testing conditions.

Source: Global Sustainable Energy Solutions

NOCT is quoted based on the following reference conditions:

- ambient air temperature of 20°C;
- irradiance of 800 W/m^2;
- air mass of 1.5; and
- wind speed of 1 m/s.

Once installed, PV modules will not perform as well as they do under STC or NOCT conditions because in 'real life' the modules are exposed to a range of losses. This may include losses due to shading, module temperature, soiling, manufacturer's tolerance and many more factors. More information on these losses and the performance of modules is given in Chapters 3 and 4.

The expected lifespan of a PV module is typically >25 years. Throughout this time, the module will be subjected to numerous different internal and external factors that could affect its actual lifespan. Some of these factors apply to all PV modules and some will apply to module types based on their manufacture. These factors include:

- weather exposure;
- yellowing (UV and heat induced);
- microfractures;
- hot spots; and
- potential induced degradation (PID).

PV modules will have manufacturers' and statutory warranties that are applicable for modules purchased. Some of the PV modules' performance losses may be able to be claimed under these warranties.

Combining PV modules

The modular nature of PV systems allows for great flexibility in system design. PV cells are connected in series to form a module, PV modules are connected in series to form a string, and PV strings are connected in parallel to form an array (Figure 2.10).

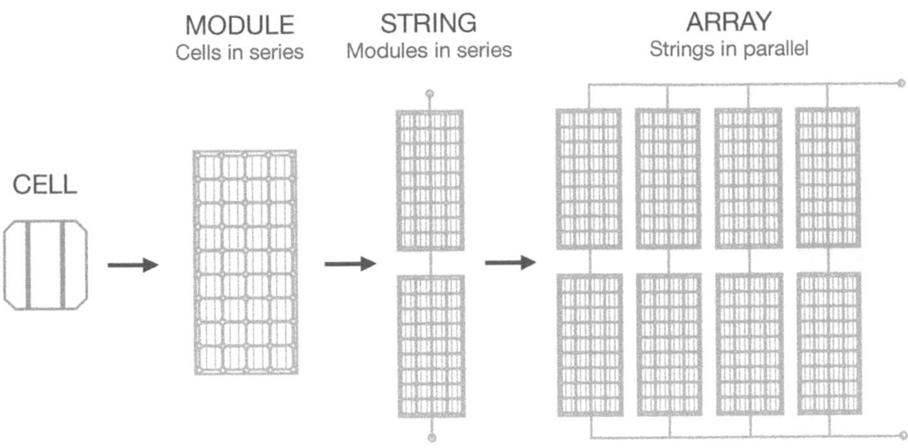

Figure 2.10 The modular arrangement of a PV array.

Source: Global Sustainable Energy Solutions

String (modules in series):

- Voltage = sum of module voltage outputs;
- Current = smallest module current output.

Array (strings in parallel):

- Voltage = lowest string voltage output;
- Current = sum of string current outputs.

For an array made up of identical modules and the same number of modules in each string, the current and voltage characteristics can be summarised as:

- Voltage = module voltage × number of modules in a string;
- Current = module current × number of strings.

Types of PV module technology

When deciding the type of module to use in a solar farm, it is important to understand the relative advantages and disadvantages of the different technologies. The two main types of solar technology are silicon crystalline cells, which may be monocrystalline or polycrystalline, and thin-film cells.

Monocrystalline cells

Monocrystalline cells are created from a single crystal of semiconductor-grade silicon, drawn out slowly from molten silicon. Although this gives it the highest efficiency, it is also the most expensive to produce. The crystal is sliced into thin wafers (0.2–0.4 mm thick) and etched to improve light trapping. An anti-reflective coating is added to the cell, making these cells dark and uniform in colour (see Figure 2.11). The cells are cut to a square shape without corners due to the way they are manufactured. Although these cells are relatively more

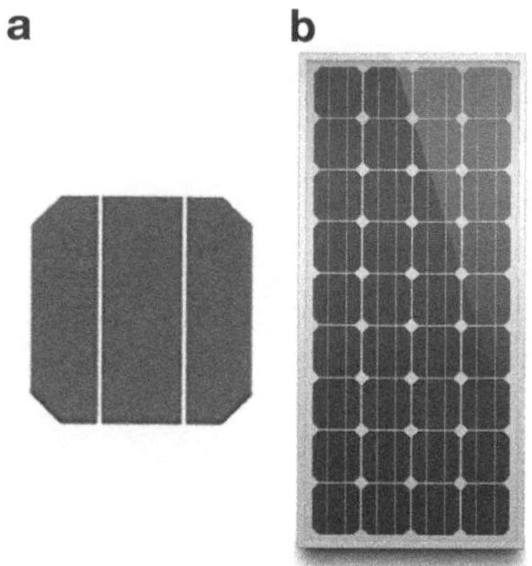

Figure 2.11 A typical monocrystalline wafer cut to a square without the corners (a) and module (b). Monocrystalline modules (c) are used on single-axis trackers in Nellis Solar Power Plant (Nevada, USA).

Sources: (a) and (b) Global Sustainable Energy Solutions, (c) enerG Magazine

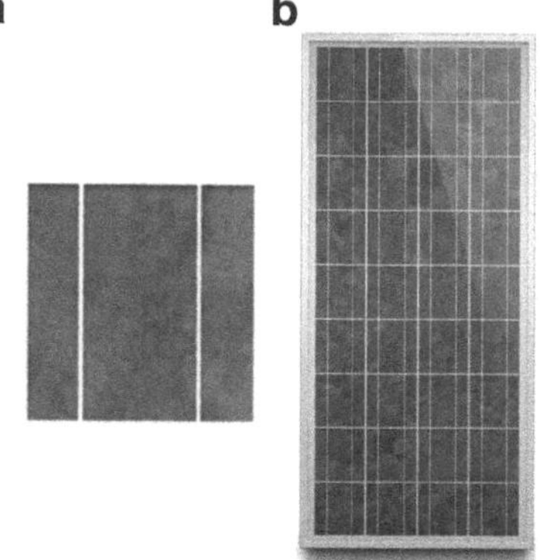

Figure 2.12 A typical polycrystalline wafer (a) and module (b). Polycrystalline modules (c) used in Dunayskaya Solar Power Station, Ukraine.

Sources: (a) and (b) Global Sustainable Energy Solutions, (c) Activ Solar

expensive to produce compared to other types of cells, the higher efficiency means that less modules, and hence less area is required for a given array power.

Polycrystalline cells

Polycrystalline cells are made up of many small crystals of semiconductor-grade silicon, creating a distinctive colour variation in the cell (see Figure 2.12). Apart from that, the manufacturing process is very similar to monocrystalline cells. Smaller crystals are easier to grow than larger ones, hence these cells are cheaper to produce than monocrystalline cells, however they have lower efficiencies. These days the prices of both technologies are very similar due to the rapid increase in global demand and the subsequent increase in the manufacture of both types of cells.

Thin-film cells

Thin-film cells are made by applying photoactive semiconductors onto a substrate (such as glass, metal or plastic) in thin layers (1–10 μm). Because they are so thin, they use much less silicon or other material to produce than

a

b

Figure 2.13 (a) Thin-film modules are not made from wafers, but from thin layers of photoactive semiconductors. (b) Thin-film modules used in one of the world's largest solar farms: Desert Sunlight Solar Farm (California, USA).

Sources: (a) Global Sustainable Energy Solutions; (b) First Solar

crystalline cells. Different semiconductor materials can be used, resulting in a few different types of thin-film cells:

- amorphous silicon (a-Si)
- cadmium telluride (CdTe)
- copper indium gallium diselinide (CIGS).

While thin-film cells are the cheapest to produce, they are also the least efficient and are subject to significant performance degradation within the first few months of use. They have lower temperature degradation coefficients, which, in some circumstances, may make them a preferable option for very hot locations. Because thin-film modules are less efficient, the quantity of thin-film modules required to reach a given installed capacity will be greater than conventional silicon products. Having more modules means more installation hardware and more labour: which impacts on the economics of the system. Cadmium-telluride thin-film modules also require specialist disposal facilities because of the toxicity of the material if released into the environment. The use of these different types of modules is further discussed in Chapter 4.

Module efficiency

Module efficiency refers to how much electricity is generated from the module compared to how much incident radiation it receives. Efficiencies that have been recorded in the lab are significantly higher than that for commercial modules due to cost limitations and optimal operating conditions in the lab. The US National Renewable Energy Laboratory (NREL) has compared the best research-cell efficiencies for different types of cells including crystalline and thin-film technologies as well as multi-junction cells and emerging PV technologies like quantum dot cells (Figure 2.14). While efficiencies of 46%

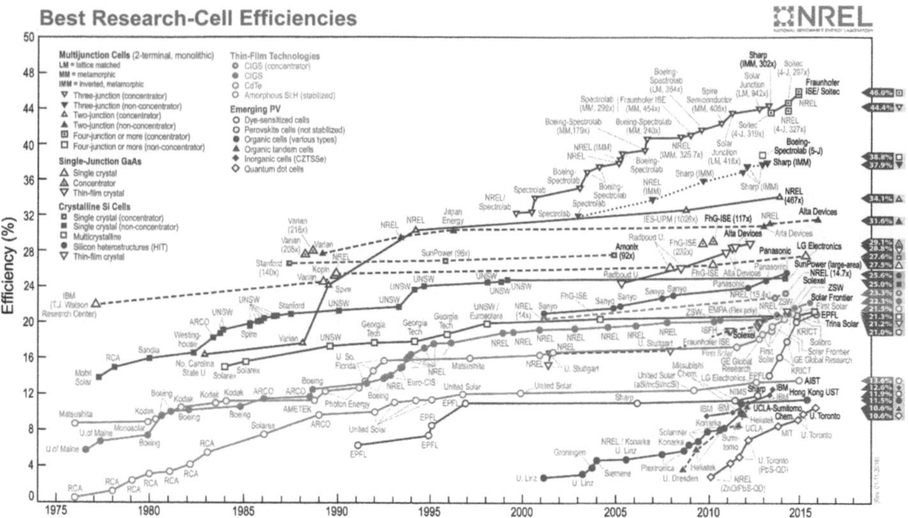

Figure 2.14 Plot of best research-cell efficiencies (under laboratory conditions).

Source: This plot is courtesy of the National Renewable Energy Laboratory, Golden, Colorado, USA

Table 2.1 Comparison of different commercial solar technologies

Cell material	Module efficiency (%)	Surface area required for 1 kWp (m²)
Monocrystalline silicon	14–20	6–7
Polycrystalline silicon	13–15	6.5–8.5
Amorphous silicon thin film	6–9	11–16.5
Cadmium-telluride thin film	9–11	9–11
Copper indium gallium diselinide thin film	10–12	8.5–10

(under laboratory conditions) can be achieved with multi-junction cells (four-junction or more), they are too expensive to be use commercially and they are used with a concentrator.

A comparison of the efficiencies of commercially available PV cell technologies is given in Table 2.1. The problem with modules that have lower efficiencies is that they require a larger surface area for a given installed capacity of PV. This means that they may be the most suitable option if there is sufficient land area available. However, having to install more modules means more foundations, more mounting structures and more time required for installation: all these additional outlays increase the cost of the project.

Regardless of which type of technology is chosen, it is important to ensure that the modules have been manufactured to a high quality and in accordance with relevant standards. These standards originate from the International Electrotechnical Commission (IEC) (www.iec.ch) and are as follows:

- IEC 61215 Crystalline silicon terrestrial photovoltaic (PV) modules – Design qualification and type approval.
- IEC 61646 Crystalline thin-film terrestrial photovoltaic (PV) modules – Design qualification and type approval.
- IEC 61730 Photovoltaic (PV) module safety qualification – requirements for construction and requirements for testing.

Inverters

Inverters, also termed 'power conditioning equipment' in some international standards documents, are a very important component of the PV system. They convert the DC electricity produced by PV modules into AC electricity that can be transported more efficiently and used to power residential, commercial and industrial loads.

The inverters used in utility-scale systems are grid-connect inverters that are capable of producing AC electricity with voltage and frequency that are compatible with the AC grid, and comply with the relevant local standards. Grid-connect inverters have integrated maximum power point trackers (MPPTs) so that the PV modules/arrays operate at their maximum MPP. The inverter must be able to reference the grid in order to connect to it otherwise it will not operate. It is a requirement for these inverters to have grid protection to ensure that they will shut down under abnormal conditions.

Grid-connect inverters differ under the following specifications:

- whether or not they have a transformer;
- the switching frequency of the transformer;
- the interface between the inverter and the PV array; and
- whether the inverter is connected at the modular, string or sub-array level.

Box 2.1 Voltage classification according to IEC 60038

The electricity distribution system applies different voltage levels according to power transmission levels required: these voltage levels are classified in various ways. For this publication, voltage levels will be classified according to IEC 60038. For reference, this is as follows:

Voltage class	Voltage range (AC)
Low voltage (LV)	< 1000 V
Medium voltage (MV)	1000 V–35 kV
High voltage (HV)	35 kV–230 kV
Extra high voltage (EHV)	> 230 kV

Inverters with and without transformers

Inverters may be transformerless or have transformers inbuilt. On the MV scale, a standard unipolar inverter may incorporate a low-voltage transformer to output the required AC voltage and provide galvanic isolation between the AC and DC side of the inverter. However, an additional transformer is still required to step the voltage up to MV levels for transmission to the substation. A transformerless inverter directly couples the AC output of the inverter to the transformer, so does not need an inbuilt transformer. Transformerless inverters are smaller and lighter, have higher efficiencies and may be clustered together before connecting to a single MV transformer, allowing for greater flexibility in design. However, in order to provide galvanic isolation, additional protective equipment must be installed to prevent DC injection to the AC grid. This may include an isolating transformer, separate secondary windings from the transformer on each inverter, DC sensitive earth-leakage circuit breakers (CB), and protection of live parts. Transformerless inverters may also allow leakage currents which cause electromagnetic interference (EMI).

Micro-inverters

Micro-inverters are small transformerless inverters that are connected to every module or every second module in an array. While very uncommon at the utility-scale level, micro-inverters have some significant advantages over other types of inverters:

- Allows each module to operate at its MPP, resulting in a higher overall output from the array.
- Less DC cabling is required, which means lower DC losses and lower installation, cabling and equipment costs, as DC protection devices are not required.
- Modularity makes it easy to add extra modules if required.
- Module-level monitoring makes troubleshooting easy and accurate.

a

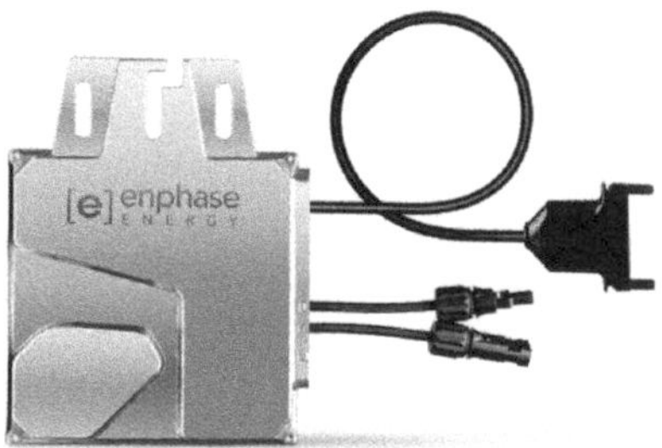

b

Figure 2.15 (a) Enphase micro-inverter. (b) Almost 2,000 Enphase micro-inverters are used on this 660 kW ground-mounted dual-axis tracking solar system in Wisconsin, USA.

Sources: (a) Enphase; (b) Convergence Energy

Disadvantages include:

- High cost per unit of power ($/W).
- Subject to high temperatures, and hence heat stress, as they are located on the modules themselves.
- If replacement is required, the whole module must be removed from the array while the inverter is serviced.

String inverters

String inverters are connected to one or to multiple strings in an array. While not as common at central inverters, string inverters are sometimes used for utility-scale application. Advantages include:

- Allows each string to operate at its MPP. This is useful if some module strings must be installed at different orientations, or if there are shading issues.
- Lower capacity compared to central inverters means more flexibility in design, and increased redundancy.
- Less DC cabling required than for central inverters.

Disadvantages include:

- High cost per unit of power ($/W) compared to central inverters.
- Increased system components compared to central inverters.

Central inverters

Central inverters are similar to string inverters, except they are connected to a larger number of strings, known as a sub-array. Therefore, using central

b

Figure 2.16 (a) SMA Sunny Mini Central 11000TL single-phase inverter. (b) This 6 MWp system in Korat, Thailand, uses 540 of these string inverters.

Source: SMA Solar Technology

a

b

Figure 2.17 (a) Sunny Central 500CP XT/630CP inverter. (b) 27.5 MWp (AC) Heideblick Solar Farm at Gahro Military Airfield, Germany, uses one 500CP XT, four 630CP XT, eighteen 760CP XT and eight 800CP XT inverters.

Source: SMA Solar Technology

inverters means that fewer inverters are required for the system. This has a number of advantages including:

- low cost per unit of power ($/W);
- high reliability; and
- simplicity of installation.

Disadvantages:

- can only track the MPP of the whole array or sub-array, unless solar optimisers are installed on the modular level; and
- no redundancy if inverter fails.

Reactive power control

Reactive power management is critical for maintaining adequate voltage levels in the electricity network. When voltage levels are too low, reactive power can be generated to raise the voltage, or reactive power can be absorbed to lower the voltage level when it is too high. Depending on the local standards and grid requirements, utility-scale PV plants must generally include reactive power control capabilities of 0.95 or 0.90 lag-to-lead power factor at the point of interconnection.

At the time of writing most utility-scale inverters are capable of monitoring and interacting with the grid to provide reactive power support. Depending on the direction of the reactive power flow from the grid, the inverter can increase its apparent current output to either increase or decrease the grid voltage. While this additional current does not affect the amount of electricity generated from the PV array, when the inverter is not operating at unity power factor, there is a slight loss in operating efficiency (approximately 0.1%). While the benefits of reactive power control outweigh this small loss in efficiency, it means that manufacturers must consider this in their designs. Manufacturers may choose to make the maximum current capacity available for real power, or have a higher current rating to leave some reserve reactive capability when operating at its rated real power. For example, the SMA inverter given in Table 2.2 has the same real (kW) and apparent (kVA) power ratings, which means that when operating at peak output it is actually only capable of producing 720 kW at a power factor of 0.9 (800 kVA × 0.9 PF). On the other hand, the AESE inverter has a higher apparent than real power rating. This means that when it is operating at full capacity it is capable of delivering close to its real power output at a power factor of 0.9: 1100 kVA × 0.9 = 990 kW. It is important for the system designers to understand these different ratings so that the loss of real power can be accounted for when modelling system yield and performance.

Protection systems

According to IEEE 1547, all grid-connected inverters must adjust their output or disconnect from the grid entirely if the AC line voltage or frequency exceeds or is less than the limits imposed by the local gird requirements. Depending on these requirements, the inverter will also be required to shut down if it detects

Table 2.2 Comparison of the reactive power control specification for different inverter models

Manufacturer	Model	kW rating	kVA rating	PF range at kW rating	Full PF range	Reactive capability at night	Real power dependent PF	Inverter-level voltage control	Var ramp rate control
AESE	AE 1000NX	1,000	1,100	0.91 lag to 0.91 lead	0.80 lag to 0.80 lead	N	Y	N	Y
Bonfiglioli	RPS TL-UL 1000	1,200	1,333	0.90 lag to 0.90 lead	0.85 lag to 0.85 lead	N	Y	Y	Y
Eaton	Power Xpert Solar 1670kW	1,670	1,830	0.91 lag to 0.91 lead	0.91 lag to 0.91 lead	N	N	N	Y
Schneider	XC 680-NA	680	680	NA	0.80 lag to 0.80 lead	Y	Y	N	N
SMA	SMA SC 800CP-US	800	800	NA	0.80 lag to 0.80 lead	coming soon	Y	Y	Y
Solectria	SGI 750XTM	750	750	NA	0.80 lag to 0.80 lead	N	Y	Y	Y

Source: Adapted from Solar Pro

that the grid is no longer present. This stops the PV system injecting voltage or current into a disconnected grid, protecting personnel who may be working on the wires or switchgear. In either of these circumstances, the inverter should remain disconnected until the grid voltage and frequency are within the appropriate limit for a period of five minutes.

The inverter should also have a means of ground fault and insulation monitoring.

DC/DC converters

DC/DC converters (also known as solar/power optimisers or 'module MPPTs') are devices which are used to ensure that each module operates at its maximum MPP. They are connected to or embedded with each PV module. While a string or central inverter MPPT will ensure that module strings or sub-arrays operate at their maximum MPP, these ensure that each module also does so. This reduces module mismatch and hence results in increased yield for the system. Therefore they can be used instead of a micro-inverter while providing many of the same benefits at a reduced system cost. However, these will increase the cost of the system. The cost of capital should be weighed up against the increased yield and hence revenue through reduced mismatch losses before deciding whether or not to use DC/DC converters in a utility-scale system.

DC/DC converters are typically built with data logging and communication functionality; manufacturers usually offer compatible inverters to communicate with these converters. While some models of DC/DC converters can function with different brands of inverters, it is important to check compatibility and functionality before selecting this equipment.

Other balance of systems components

Transformers

There will typically be two different types of transformers (other than transformers built into inverters) used in a utility-scale PV plant: the distribution transformer installed after each inverter, and a substation transformer used to step up the voltage for transmission. The primary role of these transformers is to step up the inverter output voltage to AC grid voltage levels. This could range between 12–115 kV depending on the grid connection point and its requirements. Transformers also provide galvanic isolation between the solar farm and the utility grid, which improves safety and equipment protection by preventing ground fault loops. Depending on the country of installation, medium sized PV installations may require transformers to step the voltage up to e.g. 34.5 kV as required for North American distribution voltages.

Naturally the use of any transformer will impose technical requirements for its installation and housing, and this will have associated costs. For example, a single transformer may be used to combine the output of two inverters: this application would combine the two inverter-windings into a single enclosure and output. Using transformers in this way requires the system engineers to ensure that all aspects of the transformer sizing, its cabling and temperature derating for the enclosure in question have been confirmed.

The use of any transformer will depend on the system's variables being known and therefore the transformer will typically have to be customised to ensure it will function with a specific system.

Distribution transformers

A distribution transformer is typically located after each inverter to step the voltage up from LV to MV levels for transmission to the substation and is characterised by the following:

- Pad-mounted transformers are more common than conventional pole-mounted transformers in utility-scale PV plants; this is because cabling typically runs underground rather than overhead.
- Capacity can vary between 50 to 2,500 kVA.
- Maximum voltage for pad-mounted transformers is typically 35 kV or 36 kV, however some manufacturers offer higher voltage ratings on request.
- Liquid-type transformers are almost always used instead of dry-type transformers because they dissipate heat more efficiently, require less conductor insulation and are cheaper.

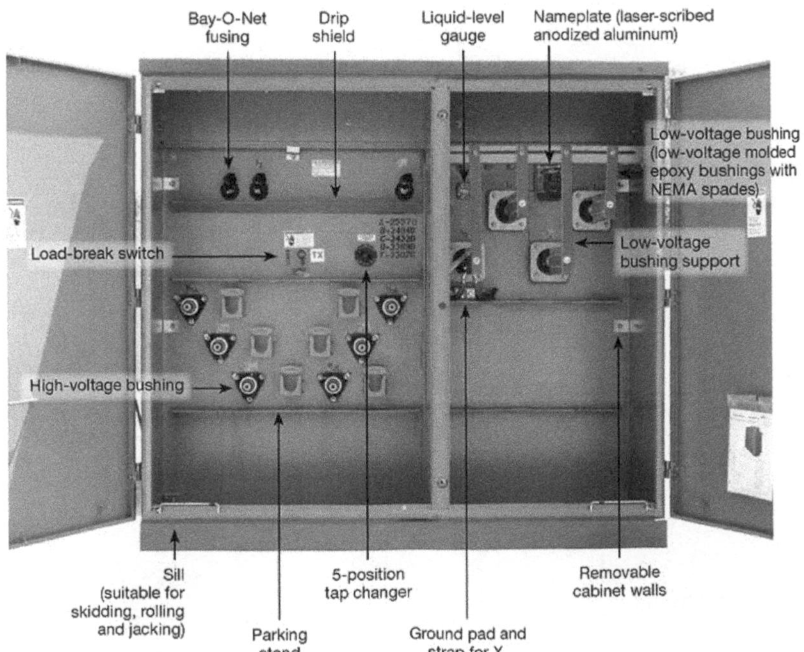

Figure 2.18 Basic anatomy of a 3-phase pad-mounted distribution transformer. The compartment on the right is the low voltage side, the left is the high voltage compartment.

Source: Image courtesy of Eaton

Substation transformers

A substation transformer is located at the substation and steps voltage up from MV to HV levels for transmission across long distances to where the electricity is distributed. This transformer is characterised by the following:

- Required when a transmission voltage is required greater than 36 kV.
- Capacity can vary from 2,500 kVA to more than 100 MVA.
- These are larger and more complex than distribution transformers and require more insulation to accommodate higher voltages.
- Liquid-cooling is used, similarly to distribution transformers.

Cabling

Cabling in a utility-scale solar plant is made up of:

- DC wiring to connect each component to the point of connection with the inverter; and
- AC wiring to connect each component from the inverter to the substation.

It is important that all cables are sized and labelled correctly and meet all applicable standards. DC cables need to be double insulated, moisture resistant and UV-resistant or installed within trunking or conduit (see IEC 61386 Conduit systems for cable management). The voltage drop on the DC side and voltage

Figure 2.19 Substation transformer along with switching and protective equipment.

Source: PEO ACWA

rise on the AC side must be accounted for. More information on voltage drop/rise is described in Chapter 4.

DC cabling will include:

- PV module cables: these are typically preconnected to the module and connect a set of PV modules in series, forming a string.
- PV string cables: these connect a string of modules to the PV string combiner box.
- PV sub-array cables: these connect the PV string combiner box to the PV array combiner box.
- PV array cables: these connect the PV array combiner box to the PV array DC switch-disconnectors.
- Inverter DC cables: these connect the PV array DC switch-disconnectors to the DC side of the inverter. This cabling may also be referred to as PV array cabling, but it has been separated in this publication to provide an extra description.

AC cabling will include:

- Inverter AC cables: these connect the AC side of the inverter to the LV/MV transformer.
- LV/MV transformer cables: these connect the LV/MV transformers to the MV collection switchgear, or substation.
- MV switchgear cables: these connect the MV switchgear to the MV/HV transformers, which connect the system to the grid.

Copper vs aluminium cables

Copper cables are more commonly used than aluminium cables as copper has a greater conductivity than aluminium, hence the copper cables can be smaller than aluminium cables to carry the same amount of current. Copper is also more ductile than aluminium so is less prone to weakening from bending during installation.

On the other hand, aluminium is cheaper than copper (about half the price per tonne). It is also about 60% lighter than copper of equal current-carrying capacity, which means that it is often more suitable for overhead lines, as large, heavy copper conductors are more complex and expensive to install overhead. For these reasons, aluminium is often more suitable when carrying larger currents.

Things to consider if using aluminium cables:

- oxidation of aluminium reduces conductivity, meaning that the oxide layer must be broken or physically removed before contacts are made;
- bi-metallic cable lugs or connectors must be used to avoid corrosion when the aluminium comes into contact with copper;
- aluminium tends to stretch and expand over time, especially when subject to higher pressure or temperature. Special crimp connections should be used to account for this expansion;

- aluminium is generally only allowed on ground-mounted arrays; and
- aluminium cables should only be used where codes specifically allow their use.

Electrical protection devices

Electrical protection devices are essential to ensure that a utility-scale solar system operates safely and is safe to access by all related personnel. Electrical protection may include overcurrent protection devices, disconnection devices, earthing/grounding and lightning/surge protection.

Overcurrent protection devices

These devices automatically disconnect during a fault and are designed to prevent damage to components and cables due to overload currents or short circuits. These include fuses which are sized so that they carry the required load current but will create an open circuit under a fault condition – the conductor material melts under excessive current, breaking the circuit; and circuit breakers which are mechanical devices installed to protect a circuit by opening that circuit under fault conditions. It can be reset when the fault is removed, so can be used repeatedly, unlike fuses.

Disconnection devices

These devices are manual switches used to break a circuit to allow parts of the system to be electrically isolated. For a utility-scale system, there should be a means of disconnection for both the DC and the AC circuits.

DC disconnection devices must be non-polarised, meaning that they operate effectively when installed in either direction. The DC disconnection devices installed in a utility-scale PV system are:

- string disconnectors – non-load-breaking disconnection switches;
- sub-array disconnectors – load-breaking disconnection switches (may be non-load-breaking); and
- PV array DC switch-disconnectors – load-breaking disconnection switches.

On the AC side:

- inverter disconnectors – load-breaking disconnection switches (may be non-load-breaking);
- main switchboard disconnectors – load-breaking disconnection switches; and
- transformer disconnectors – load-breaking disconnection switches.

Earthing/grounding

This protects both the system equipment and people from dangerous fault conditions by directing fault current to earth via an earth stake. Earthing can also be used to improve the performance in some systems.

Equipotential bonding (also known as protective earthing) is used between exposed conductive parts of the PV array to ensure that there is no voltage (potential) difference between any two components in order to protect people from electric shock. Functional earthing is only necessary for certain types or brands of modules to ensure optimal performance of the array.

Lightning/surge protection

This involves a combination of earthing/grounding and surge protection devices (SPDs) in the event of nearby lightning strikes. Lightning/surge protection systems are designed to protect equipment from being directly struck by lightning, or being in the path of discharge from a strike, by directing the strike through large current-carrying conductors to earth. SPDs differ in that they provide overvoltage protection for direct and indirect lightning strikes causing magnetic fields that induce transient currents. The associated transient voltages appear at equipment terminals and may cause insulation and dielectric failures to important system components.

Appropriately rated surge protection devices (SPDs) are usually installed as close as possible to the inverter in the combiner box on the DC side to protect from strikes on the array, and the AC side to protect from strikes on the grid. SPDs should also be installed on the grid side of the AC switchgear.

Junction/combiner boxes

For a utility-scale solar plant, a string junction/combiner box will be used to interconnect the string cables of each sub-array connected to the inverter input. In some cases, an array combiner box may be used to house the connection of these sub-arrays before entering the inverter. However, most utility-scale central inverters have multiple DC inputs, so the sub-arrays are paralleled inside the inverter, leaving no need for the array combiner box. Junction boxes may also be required at points where AC cables from inverters are connected together and then connected to the substation via one AC cable. Where necessary, junction boxes will also house the overcurrent protection, the disconnection devices, the SPDs and monitoring equipment.

Metering

Unlike grid-connected rooftop installations, solar farms export the vast majority of generation to the electricity grid, and consume a small percentage on-site (the auxiliary load). It is essential for a solar farm to measure the electricity generated and exported to the grid so that revenue can be calculated based on the reading. A revenue type interval meter should be installed on the grid side of the transformer to measure the amount of electricity exported to the grid in order to calculate revenue from feed-in tariffs, power purchase agreements (PPAs) or the spot market, as well as providing a means of monitoring system performance. An import–export revenue type interval meter should also be installed at the low voltage switchboard on the auxiliary circuit to measure the electricity consumed by the auxiliary loads at the solar farm. The meter measures electricity

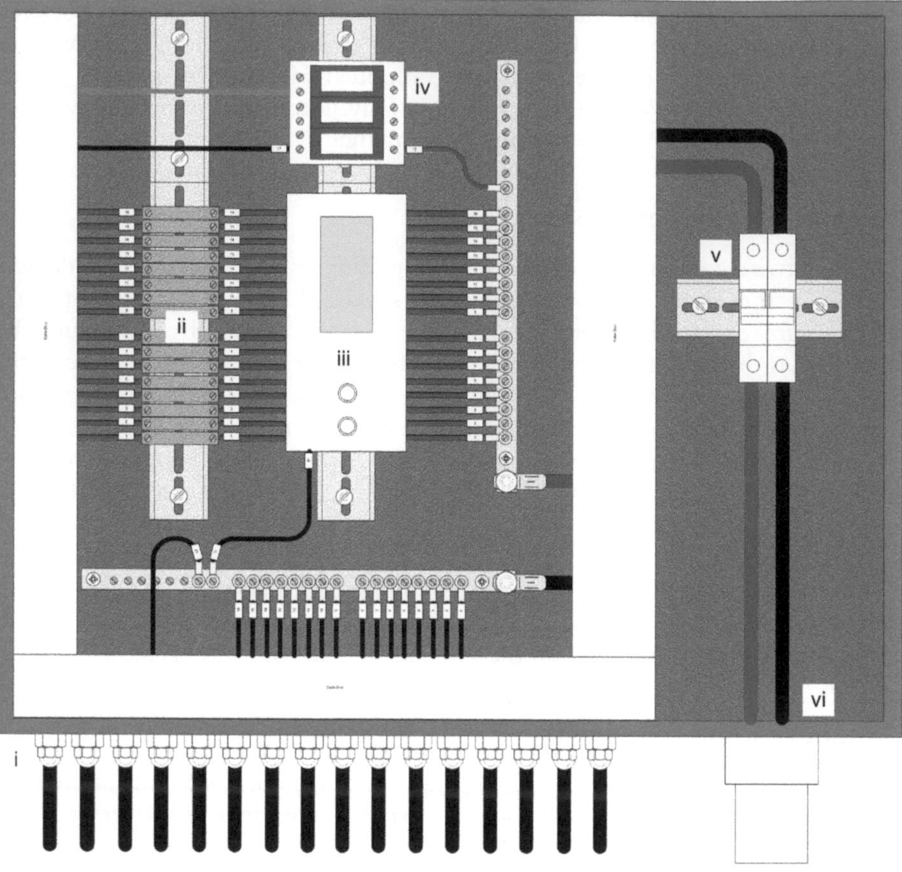

Figure 2.20 A string combiner box for a 90 MW solar farm in South Africa. It has 16 inlets, with string protection and monitoring equipment. The 16 positive and negative string cables go through the base of the combiner box (i). The negative (black) cables are combined in their respective circuit breakers (string overcurrent protection) (ii), while the positive (red) cables pass through the string monitor (iii) before they are combined. The sub-array cables then go through the surge protection (iv) and the array DC switch-disconnector (v) before going out of the base to the inverter or array combiner box (vi).

generation from the solar farm. All metering devices should comply with local interconnection codes and standards.

System monitoring

Utility-scale solar systems require a high level of monitoring and control to ensure that the system runs optimally and in accordance with the local grid requirements. Supervisory control and data acquisition (SCADA) systems are used to monitor and control various components of the PV system. There are different methods and equipment used to achieve SCADA, but the goal is generally the same (see Figure 2.21).

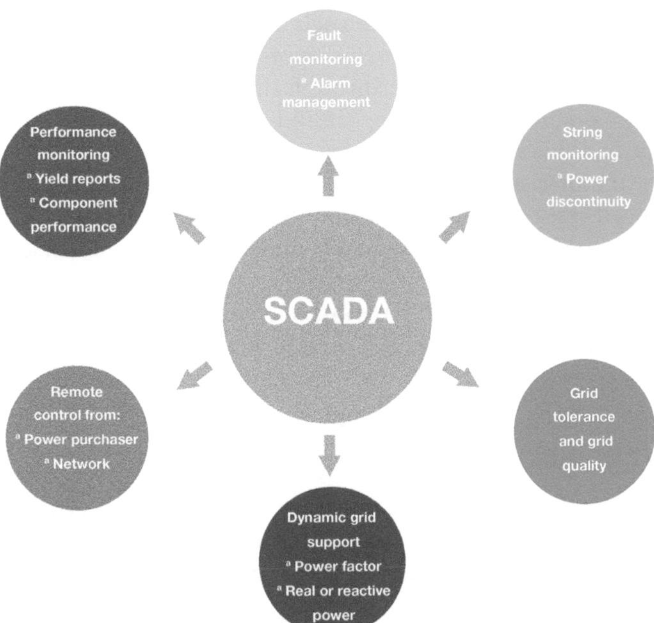

Figure 2.21 SCADA system functionality in a solar plant.

The following equipment can be used in the provision of SCADA services for utility-scale PV systems:

- programmable logic controllers (PLCs);
- ripple control receivers;
- demand response-enabled devices (DREDs);
- multifunction protection relays;
- object linking and embedding (OLE) for process control (OPC server);
- power line communication;
- serial/Ethernet/Wi-Fi monitoring;
- weather sensors and weather prediction algorithms; and
- building management systems (BMS) and energy management systems (EMS).

For more information on SCADA and system monitoring, see Chapter 6.

Quality of components

Product testing

Product failures on-site affect a number of parties including: the developer incurring a loss of system yield; installers having to provide costly callbacks; and manufacturers having to support the cost of liability claims. Testing products to a universally accepted standard is critical for ensuring product reliability and safety compliance. For example, problems causing degradation and reduction in module performance such as broken interconnects, moisture ingress, delamination, micro-cracks, hotspots, ground faults and structural failures can be identified in advance through thorough product testing. For PV modules, design qualification testing should adhere to IEC 61215 and IEC 61646 standards, however some industry experts believe that these standards only represent the first six to eight years of module life in the field. Moreover, industry standard tests do not predict the lifetime performance or degradation in the field under varying conditions and climates. Therefore, it can be very beneficial when selecting system equipment (PV modules, inverters, transformers, etc.) to choose manufacturers that have not only adhered to international testing standards but have implemented tougher requirements such as specific tests for operation in certain climates.

Certification and safety marks

Before any products can make it to the market, they must be certified to ensure the product's electrical safety according to national standards. The IEC issues internationally accepted standards for PV modules, inverters, transformers, switchgear and wiring, etc., but different countries will apply different requirements at a national, regional and local level, all of which must be adhered to. There is a large range of electrical safety approval marks that apply to different countries' standards and electrical safety requirements, so it is important to select products with the appropriate safety marks. Some examples of different safety marks used in different countries are given in Figure 2.22.

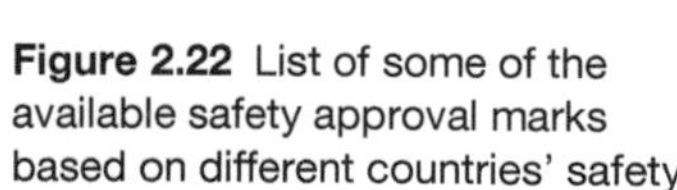

Figure 2.22 List of some of the available safety approval marks based on different countries' safety standards.

PV system design criteria

The design of a utility-scale solar system should incorporate various criteria identified as necessary to meet the system's performance outcomes. These outcomes may include but are not limited to:

- Maximise safety
  - Design the system to minimise hazards to people on-site, and those responsible for installing and maintaining the system.
  - Wiring design and electrical protection devices should be designed in accordance with national and local codes and standards.
- Minimise initial investment
  - The design will be limited by the amount of capital available.
  - The design should attempt to minimise capital expenditure as much as possible without compromising the quality of electrical equipment, incurring high operating costs.
- Minimise downtime
  - To ensure profitability of the solar farm, the system should be designed to minimise the amount of time it is not operational.
  - This can be done using high quality electrical equipment and conductors; using best installation methods; designing appropriate system alarms, monitoring and diagnostics; and selecting preventative maintenance systems or equipment that sends an alarm before an outage occurs.
- Maximise flexibility and expandability
  - Design the system so that it can be adjusted for future changes or expanded.
  - Using smaller inverters rather than fewer large inverters may increase the system flexibility.
- Maximise electrical efficiency and minimise operating costs
  - Design the plant to minimise cable, PV, inverter and transformer losses.
  - Higher efficiency usually means higher capital costs (but lower operating costs); therefore, a balance must be made between the capital available, and the cost of future money.
- Minimize maintenance costs
  - Typically, the more complicated the system, the more maintenance will be required; e.g. solar farms using trackers will require more maintenance than stationary mounted structures.
  - If the system design is kept as simple as possible and equipment is of high quality, maintenance costs will be reduced.
  - Regular maintenance will reduce the need for expensive fixes or replacement of equipment.
- Maximize power quality
  - The design should ensure that the voltage, frequency and power factor injected to the grid comply with the local grid requirements.
- Ensure compatibility of equipment
  - It is important to select appropriately sized equipment that is compatible with the equipment it interacts with.
  - Consult manufacturer's guidelines.

Bibliography

Bozicevich, R. 'Solar Module Testing Practices: Future Standards, Current Limitations'. *Solar Industry*, 2013, 2 edn.

Brucke, P. 'Reactive Power Control in Utility-Scale PV Plants'. *Solar Pro – Optimal Design, Installation & Performance*, June/July 2014.

GSES. 'Chapter 20 – Commercial and Utility Scale PV Systems'. In *Grid-Connected PV Systems: Design and Installation*, 425–6. Sydney, 2015.

GSES. 'Chapter 3 – Solar Radiation'. In *Grid-Connected PV Systems: Design and Installation*, 27–51. Sydney, 2015.

GSES. 'Chapter 7 – Inverters'. In *Grid-Connected PV Systems: Design and Installation*, 103–37. Sydney, 2015.

GSES. 'Chapter 9 – Balance of System'. In *Grid-Connected PV Systems: Design and Installation*, 159–79. Sydney, 2015.

Helukabel. 'Aluminium as a Conductor Material: A Lighter and More Economical Alternative'. 2015.

IFC. 'Utility Scale Solar Power Plants: A Guide For Developers and Investors'. 2012.

IPP. 'Interconnection Facility Agreement'. 2015.

Kondrashov, A. and T. Booth. 'Distribution and Substation Transformers'. *Solar Pro*, 2015.

Schlesinger, R. 'Surge Protection Devices for PV Installations'. *Solar Pro – Central Inverter Trends in Power Plant Applications*, 2010 October/November: 22–7.

Worden, J. and M. Zuercher-Martinson. 'How Inverters Work'. *Solar Pro: PV Source Circuit & Array Combiners DC Side Connections*, 2009 April/May.

3

Planning a solar farm

For a solar farm to be labelled a 'success', the scope of the solar farm project must have been addressed in relation to all elements of the project: practical, financial, legal etc. As the information relating to these elements is shown to combine favourably, the solar farm project evolves.

This chapter sets out those elements of a solar farm which underpin the viability and the physical requirements of the project, i.e. is the space to be used for the solar farm suitable, large enough and available? Can the proposed site legally be used for the solar farm? Can the interconnection permits be obtained from the proposed network interface? What power production is able to be guaranteed based on industry standards from the proposed solar farm? Can the solar farm project be realised and then operated safely? What are the economics of the proposal? What stimuli can be applied at government levels? What financial metrics are used to measure the solar farm's performance?

Determine the purpose

The purpose of the solar farm must be determined and agreed upon as the first step in planning a solar farm project. The solar farm's design and plan must be optimised to meet the specified outcomes and deliverables of the project. Solar farms may be built for a variety of reasons, such as providing power to meet increased localised electricity demand or expanding electricity access; as a means to reducing emissions in accordance with national and international policies and goals; providing investment opportunities or to meet local communities' expectations based on financial and/or environmental deliverables.

The pre-eminent reasons can be broken down according to practical and financial rationales; some examples are given in Figure 3.1.

The design and planning processes should be tailored to achieve the identified overarching purpose or purposes to be derived from building the solar farm.

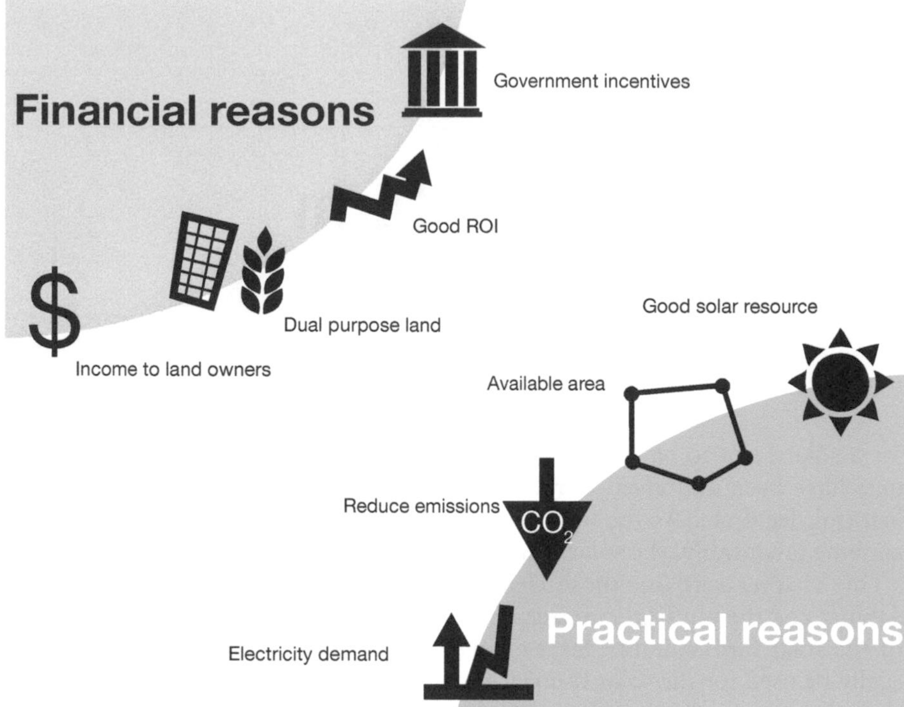

Figure 3.1 Reasons for building a solar farm.

Source: Global Sustainable Energy Solutions

Site selection

Successful project planning and solar farm implementation starts by determining a suitable site. Assessing site suitability involves assessing the solar resource, consulting stakeholders, ensuring grid connection is feasible, evaluating the topography of the land as well as many other environmental and social considerations. The main concerns involve climate (e.g. flooding, snow and fires), inadequate solar resource, soil conditions and geopolitical attributes, geotechnical attributes, current land use, and existing infrastructure including roads and transmission lines.

Key tips:

- It is important that land options in the preferred location are secured at the early stages of planning.
- Ensure site selection and due diligence are carried out thoroughly to reduce permitting and ongoing land use agreement issues.
- Interconnection of the solar system to the grid can be a major barrier. It is important to engage with network service providers and competent consultants early in the project's life to ensure grid connection of the proposed solar farm is feasible.

Site topography

The topography of the site is an important consideration and will include the shape, slope and surface features of the land in question. In order to maximise output from the solar farm, each PV module must be tilted towards the sun in a uniform direction. This can be achieved by levelling the ground with earth-moving equipment or allowing the modules to follow the topography of the land, minimising earthworks. Therefore, topography comprising sloped or undulating land, ridges, rocks, etc. can increase the complexity and hence the time and cost of the project. Figure 3.2 shows how the modules were designed to follow the undulating hills on Les Mées solar farm in France. If the site is located in a valley or close to a body of water, it may be in a flood-prone area: more details on flooding risks will be given in the section below on climate and weather. If ground levelling is required in order to build the solar farm at the site, the extent of this levelling should be confirmed and the costs estimated in the pre-application stage of planning. Although flat land is typically preferred, especially for tracking systems, sometimes alternative topographic attributes can be favourable, for example, sloped land may help to block the site from view, reducing the visual impact. Sloped land can also reduce the quantity of framing structures required if the land is sloped to face in the direction of the sun (north in the southern hemisphere, south in the northern hemisphere).

There are many digital cartographic/geographic data files available for download or purchase to gather information about the contours of proposed

Figure 3.2 Les Mées solar farm, France, has been designed to follow the contours of the undulating hills of La Colle des Mées plateau. This has not only increased the aesthetic value of the solar farm, but has also improved the amount of solar radiation collected by the modules by allowing them to be tilted in uniform direction.

Source: Christian Pinatel de Salvator

Figure 3.3 Terrain data can be analysed using HELIOS 3D, the colours indicating which parts of the site have north, east, south or west orientation.

Source: Stoer+Sauer

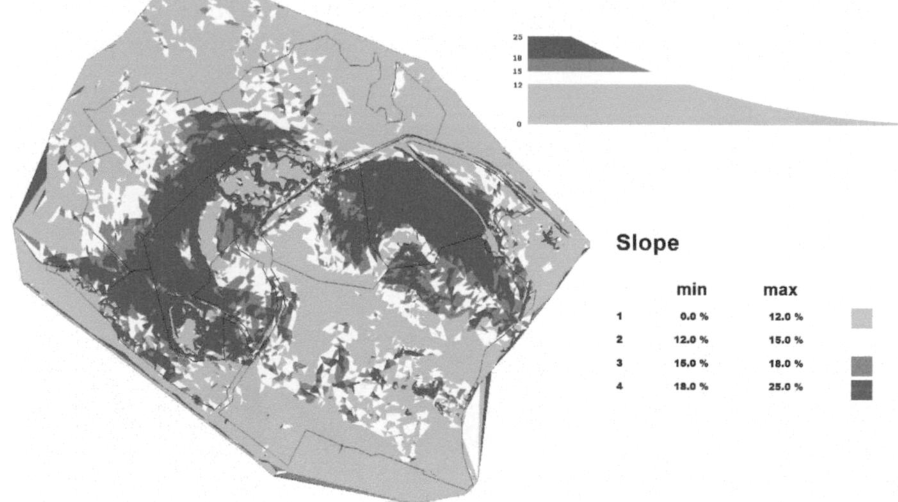

Figure 3.4 Terrain data can be analysed using HELIOS 3D to show the elevation and angle of slope to get a better understanding of the structure of the terrain.

Source: Stoer+Sauer

sites as well as surrounding features. However, during the final stages of development a detailed land survey should be completed in order to get accurate terrain data. If a recent topographic map of the site from a reliable resource is not available a suitable service provider should be contracted to prepare one. This data can then be analysed using software such as HELIOS 3D to determine orientation and slope of the surfaces on-site, and can even determine which areas are likely to erode.

Climate and weather

Solar resource

One of the most important considerations for the siting of a solar farm is the level and variability of solar irradiation at the site. Areas with consistently high levels of irradiation are favourable and give higher yield and performance. However, all proposed sites must be assessed for the location's daily and seasonal variation in solar resource. Ideal locations will have long daylight hours, minimal cloud cover and have little risk of shading/soiling from snow or dust.

Tropical areas close to the equator tend to have a better solar resource in the dry season, with longer days and clearer skies; however, during the rainy season there is frequent cloud cover, which can substantially reduce energy output of the system. Areas close to the poles tend to have very poor output during winter because of the minimal daylight hours and soiling from snow cover, but these areas will have a much higher output during the summer months. Overall, solar irradiation is better in summer months primarily due to the longer daylight hours, but it is necessary to conduct detailed analyses of the site's radiation data to gain a thorough understanding of expected performance.

Snow

Snow cover can also have detrimental effects, causing performance loss from shading or structural damage from the downward pressure of compacted snow on the modules. If a solar farm is built in an area prone to snowfall, frequent maintenance may be required during winter to remove the snow cover. The viability of frequent maintenance may have to be assessed to confirm that the costs of loss of production because of the snow covering all or part of the module face outweigh the costs associated with removing the snow. The proposed system design could require that the modules be mounted at a high tilt angle: this can also help reduce the impacts of snow and often a high tilt angle is required for locations at high latitudes as a standard design parameter. It is also necessary that the modules' mounting structures are installed high enough off the ground level so that the base of the modules are always above the expected snow level in the area.

Rainfall

Although some rainfall can be an advantage for solar farms because it acts as a cleaning mechanism, too much rainfall can reduce the overall performance because it means increased cloud cover or flooding in extreme cases. It is therefore important to assess the volume and frequency of rainfall at the prospective site. High rainfall coupled with unfavourable topography at the site can mean that the chances of flooding are high.

Flooding can cause severe damage to a solar farm by flushing and eroding the land beneath the mounting structures, as well as damaging electronic ancillary equipment, incurring significant repair costs. Flooding may also block access roads and maintenance areas, which can delay the repair and restoration of the solar farm. In order to assess and mitigate any flood risk, a risk assessment should be conducted followed by a mitigation plan if necessary (Figure 3.5).

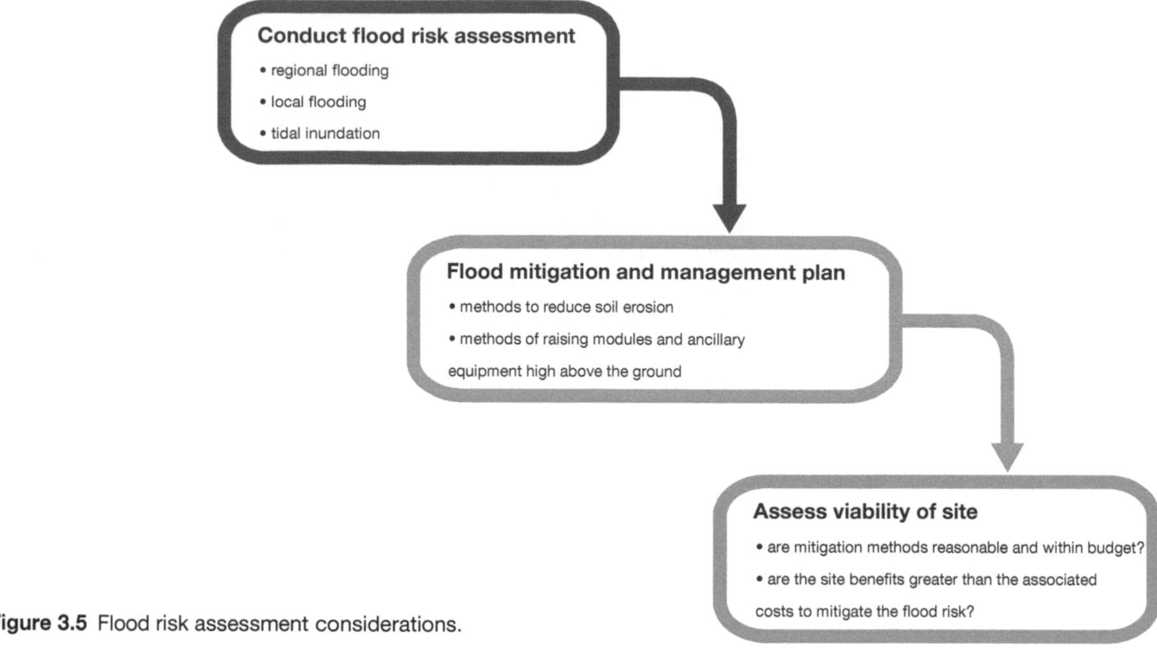

Figure 3.5 Flood risk assessment considerations.

Temperature

It is important to establish accurate temperature data for the site in order to predict the system performance. Higher temperatures usually affect the solar system's power output and therefore the system's performance. Temperature data is required to determine the system's maximum and minimum array voltages, and this information is in turn required for sizing inverters (more information on this is given in Chapter 4).

Soil type and vegetation

The type, or types, of soil at the proposed site must be confirmed before being able to determine the system design and whether or not the ground is capable of providing sufficient support for the mounting structures or foundations. Determining the soil type will also be required to identify which foundation materials are suitable for use.

A geotechnical survey of the site will determine the groundwater level, resistivity of the soil, load-bearing properties, the presence of rocks and soil pH levels (to determine the level of corrosion protection required). This survey can also provide data to determine the risk of seismic activity, flooding, erosion and frost. For example, the geotechnical investigation for the Sunshine Coast Solar Farm in Australia consisted of eight boreholes drilled to depths between 8–12 metres, six core penetration tests pushed to 8 metres, six test pits excavated to 3 metres depth as well as two soil permeability and 16 resistivity tests. The results showed that the site provided soft or unstable foundations consisting of soft clay, silt and loose sands. This meant that the site was at risk of becoming non-trafficable, which means that vehicles might not have been able to travel around the site unless sufficient drainage measures and construction practices were maintained. It was recommended that temporary working platforms be used during construction and that high-level footings should not be used without performing appropriate groundworks to improve the bearing capacity of the soil (see Figure 3.6).

Most sites will have some level of native or introduced vegetation, and it is important to assess and categorise this vegetation in order to determine what may need to be removed or what level of maintenance will be required to keep it under control. The most suitable site would have low level vegetation requiring minimal maintenance. However, a lot of sites will have long grass, trees or shrubs that can soil or shade the modules. An example of poor maintenance is given in Figure 3.7; although the spacing between the rows has been maintained, the grass under the modules has been ignored, causing significant shading and hence loss of performance from the modules. An example of a well-maintained solar farm is given in Figure 3.8.

Methods of managing vegetation include mowing, trimming, spraying or mulching. Spraying and mulching might not be the preferred options because they may affect the visual impact of the site. An acceptable alternative is to allow farm animals like sheep, chickens or geese, to graze around the solar farm. This practice means that the modules will need to be raised to a suitable

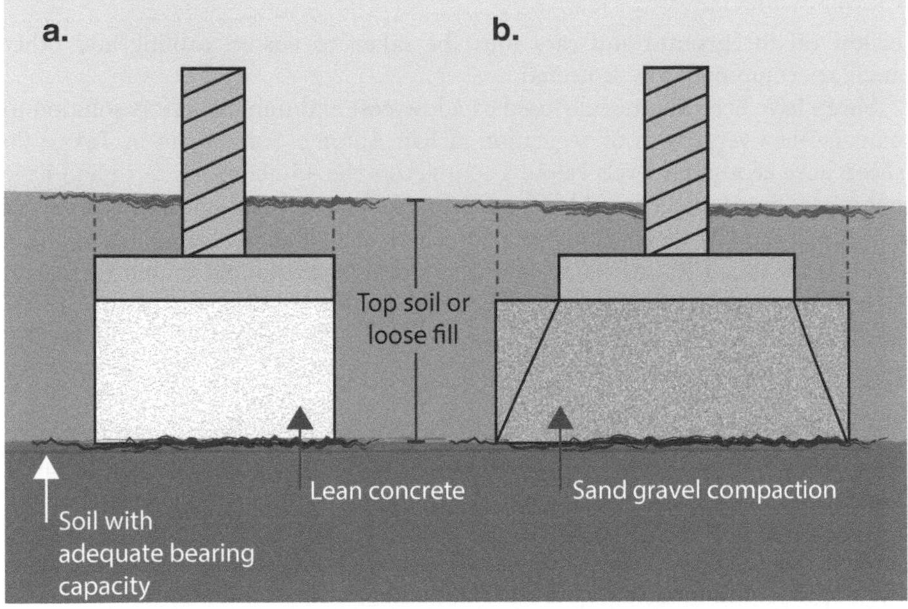

Figure 3.6 High-level footings are footings which sit just under the topsoil. However, if this soil does not have adequate bearing capacity, appropriate groundworks may include: (a) removing the topsoil directly below the footing and replace it with lean concrete; (b) removing the topsoil in an area larger than the footing and replacing it with compacted sand and gravel fill. The area of the compacted sand and gravel fill should be sufficiently large to distribute the footing load.

Figure 3.7 (a) Poorly maintained solar farm in Germany. (b) Greenough River Solar Farm, WA, Australia, is an example of a well-maintained solar farm, with grass kept short and no sign of overgrowth.

Source: (a) Greentech Media (b) First Solar

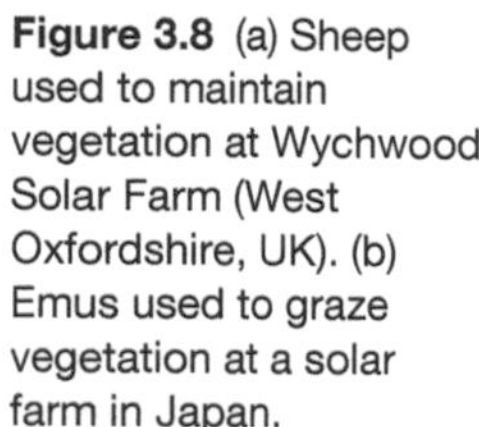

Figure 3.8 (a) Sheep used to maintain vegetation at Wychwood Solar Farm (West Oxfordshire, UK). (b) Emus used to graze vegetation at a solar farm in Japan.

Source: (a) Sam Simson; (b) Oita Sekiyu

height off the ground and care must be taken to ensure cabling and other ancillary equipment are protected.

Sheep have been successfully used as a low-cost and minimal effort solution to manage the overgrowth of vegetation at San Antonio Solar Farm in Texas: 90 sheep have kept grass levels below 1 foot across the 45-acre solar farm and have not caused any damage to cables or modules. An unusual example of this is the use of emus at Usa City Solar Farm in Japan where these animals maintain vegetation levels at the site. The emus were considered ideal because of their ability to adapt to both cold and hot weather as well as for their quiet and friendly nature.

Current land use

Since solar farms occupy large areas of land, it is important that the proposed solar farm's installation and associated works will have minimal negative impacts on current land usage and that planning remains in accordance with local government development plans. Ideally a solar farm is built on land of limited financial or community value such as land having been previously developed, contaminated or industrial land.

In the US, the Environmental Protection Agency is funding an initiative to encourage the use of contaminated land, landfills and mining sites for renewable energy projects to reduce the use of green space for development. As of 2013, this programme has regenerated 49 such sites as solar farms. Often though, the most suitable site is on agricultural land. This presents an issue if the land supports viable and diverse agriculture, meaning that the land may have more value for agricultural use rather than for a solar farm. Approval for such installations usually requires that the land is restored to its original use once the solar farm has been decommissioned. The UK National Planning Policy Framework outlines the guidelines for developers in relation to agricultural land classification, shown in Figure 3.9. If a solar farm is proposed for agricultural land, it is likely to encounter issues with approval if the land is classified as Grade 1, 2 or 3a (see Table 3.1).

Table 3.1 UK agricultural land classification.

Grade	Principal use
1	Intensive arable cropping, e.g. bulbs, vegetables, roots and cereals. No forestry
2	Arable cropping/intensive grassland e.g. cereals with roots and/or dairy cows. Limited forestry
3a & 3b	Extensive arable cropping, rotational grassland, e.g. cereals, oilseed rape and beans or grass leys for dairy cows, beef, sheep. Hardwood forestry mainly
4	Permanent grassland/rough grazing, e.g. beef and sheep rearing with limited dairying and cereals. Commercial softwood forestry
5	Rough grazing often with rock outcrops, e.g. principal summer grazing with hardy sheep breeds and hill cattle. Limited softwood forestry

Source: Lawson Fairbank

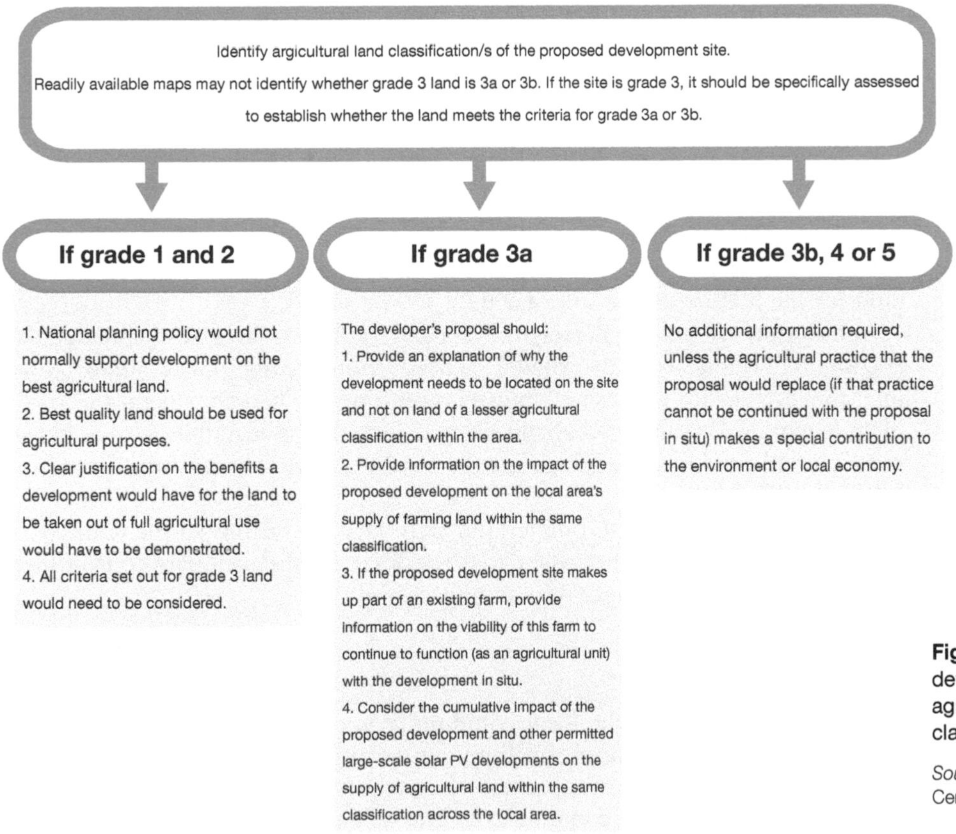

Identify agricultural land classification/s of the proposed development site.
Readily available maps may not identify whether grade 3 land is 3a or 3b. If the site is grade 3, it should be specifically assessed to establish whether the land meets the criteria for grade 3a or 3b.

If grade 1 and 2

1. National planning policy would not normally support development on the best agricultural land.
2. Best quality land should be used for agricultural purposes.
3. Clear justification on the benefits a development would have for the land to be taken out of full agricultural use would have to be demonstrated.
4. All criteria set out for grade 3 land would need to be considered.

If grade 3a

The developer's proposal should:

1. Provide an explanation of why the development needs to be located on the site and not on land of a lesser agricultural classification within the area.
2. Provide information on the impact of the proposed development on the local area's supply of farming land within the same classification.
3. If the proposed development site makes up part of an existing farm, provide information on the viability of this farm to continue to function (as an agricultural unit) with the development in situ.
4. Consider the cumulative impact of the proposed development and other permitted large-scale solar PV developments on the supply of agricultural land within the same classification across the local area.

If grade 3b, 4 or 5

No additional information required, unless the agricultural practice that the proposal would replace (if that practice cannot be continued with the proposal in situ) makes a special contribution to the environment or local economy.

Figure 3.9 Steps for developers on agricultural land classification in the UK.

Source: BRE National Solar Centre

However, since solar farms typically consist of modules set on piles, once the solar farm has been constructed, there is little ongoing disturbance from humans or machinery on the site and therefore there is an opportunity to enhance the biodiversity of the land beneath the solar farm. All sites are different and may have different habitat enhancement opportunities such as boundary features like hedgerows, grassland habitats like wildflower meadows or ponds for invertebrates, amphibians, birds and reptiles. Although these features will increase the ecological benefits of the solar farm, they require appropriate and regular management and monitoring which will incur additional operational expenses. A qualified ecologist will be needed to evaluate the various biodiversity options and to develop a management plan. The opportunity may exist to use the space for crops also, which can increase the economic benefits of the solar farm and improve the prospects of the site approval process.

Available area

When selecting a site, it is critical to determine the amount of shade-free land available and how many solar modules could physically fit at the site, including space required between rows for maintenance and to minimise inter-row shading. There must also be space for installation and maintenance vehicular access, equipment and site monitoring enclosures, fences and any other structures that may need to be built. Depending on the shape of the available area and which direction the modules should be tilted, the site may fit more modules in either landscape or portrait orientation.

Land ownership

Potential sites may be located on public (government/crown) land or may be privately owned. These ownership models will present different legal requirements to reach agreement so that the site may be procured or leased. The developer must obviously undertake to purchase the land or secure rights to the land for the lifetime of the solar farm.

Acquisition of the ownership or leasing rights to any land should also encompass rights to any roads, access routes and, most importantly, interconnection to electricity transmission lines. This consideration could include managing various challenges e.g. access or the installation may bridge various land holdings, meaning that the land rights and acquisition resolution might be difficult.

The mechanism to lease public land will vary according to the country or local jurisdiction so it is necessary that the solar farm developers consult local regulators in the early stages of planning.

Community support

It is recommended that the local community is consulted and engaged in the early stages of any proposed development: strong or reasonable objections to the project can slow or even prevent the project from reaching construction. Communication with the community is best achieved through local exhibition/

presentations where sufficient information is made available to the public, allowing the views and opinions of local people to be heard and noted so that suitable responses can be made, and hopefully these comments are taken into consideration before the final planning application is submitted. Listening to the issues that are important to the community can be a more valuable exercise than promoting the project and having the developer assume that they know what the community wants. It is important that the local community are aware of the facts of the project, including the impacts on and benefits to the local community.

Community support can be improved by the developer initiating opportunities for community gain, such as establishing a local environment trust to fund energy conservation in the local area, or to enable local or community ownership of modules etc. Greenough River Solar Farm in Australia established a community support initiative that funds scholarships to local university students studying in the field of science, engineering, technology and renewable initiatives. For more information, see the final section of this chapter on community engagement.

Regulations and permitting

Solar farms are often subject to stringent review processes through federal, state and local regulators and jurisdictions. Some important considerations for siting a solar farm include land use permits, transmission and interconnection requirements, water rights and environmental impact assessments. These can take years and consume large sums of money before a project gains approval. Understanding local regulatory requirements is key in selecting a site that will allow the project to be completed on time and on budget.

In the US, a right-of-way (ROW) grant (permission) must be obtained before any development can occur on federal land, taking between three and five years from initial application to approval. A summary of the review process is given in Figure 3.10.

Siting of transmission lines and interconnection approval can also be a lengthy process. Depending on the site location, in the US, the developer may be required to acquire a federal ROW, obtain approval from state or local governments, or

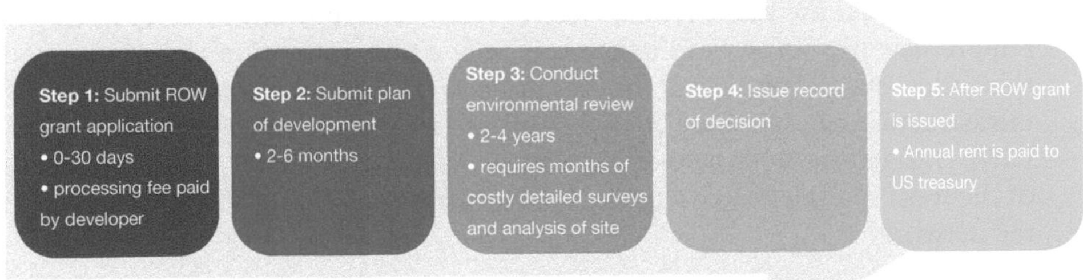

Figure 3.10 The US Bureau of Land Management permitting process for developments on public land.

Source: SEIA, 2013

go through a 'state encroachment process' (obtaining a permit to build these objects on state land). Since solar farms typically require connection to the grid, the developer must also obtain an interconnection agreement that usually involves specific requirements depending on the location.

Each site is inherently different, meaning there will be different permitting and regulatory requirements depending on the country, state, district, and location in relation to other facilities.

Fortunately, work is being done to improve these processes: the US National Renewable Energy Laboratory (NREL) is developing a regulatory and permitting information desktop (RAPID) roadmap to help optimise and streamline the regulatory process for solar farms. The RAPID roadmap is being developed, through the engagement of both US state and federal agencies, to review and coordinate the permitting process in order to reduce developers' time and costs for large-scale solar projects.

The US Interstate Renewable Energy Council (IREC) (www.irecusa.org) has developed a suite of publications addressing the raft of approvals and permissions required by US states for the integration of solar at rooftop and larger scales. The associated costs necessary to formalise an application for a solar system through to approval, i.e. the non-construction costs, have been termed 'soft costs'; and it was recognised these 'soft costs' were in fact an unacceptably large proportion of these systems' costs. IREC's publications summarise the state-based requirements and costs for solar systems' interconnection with the grid. This type and degree of information is required before any project can be assessed for practical and financial suitability.

Network and transmission lines

As outlined above, connecting a solar farm to the electricity network can prove to be a lengthy process from a regulatory point of view, but it can also be challenging and expensive to achieve the completion of the system's installation and interconnection. It is therefore critical to engage early with network service providers to verify the suitability of the site from a connection standpoint, ensuring that the solar farm will not have adverse impact on network stability and system security. If the local electrical grid cannot handle the system's proposed solar power injection, the project may not be viable for that location. Additionally, if the site is remote, any interconnection to the existing grid may require new transmission infrastructure, e.g. a new substation to be built, and this adds significant costs to the project. If possible, access to existing infrastructure can minimise costs as well as visual and environmental impacts. Early engagement with network service providers also allows for development of the most suitable reactive power control strategy for both the solar farm and the network.

Bushfire/wildfire risks

A bushfire/wildfire risk analysis should be completed when determining a site's suitability. It is important to first consult the local council to obtain a bushfire-prone land map (based on the type of vegetation in the area) followed by an on-site investigation into the potential fire risk. If the area is bush/wild fire-prone it is important to comply with development standards for this land. For example, if the land is dominated by perennial grasses, then the fire risk will be dominated by grass fires and relevant mitigation measures such as regular cutting or grazing of the site's land must be put in place. Depending on the degree to which the site is prone to bushfires, the mitigation measures may be too costly or the risk of fires may be too high to render the site a viable option. NSW Rural Fire Service in Australia recommends that a combination of bushfire protection measures, given in Figure 3.11, depending on the specific site conditions, should be implemented to achieve the lowest fire risk.

Cultural heritage sites

Generally, if the site is located within a cultural heritage listed area, a project will not usually receive development application approval, as these sites are protected from development due to their cultural significance. It may be necessary to conduct a cultural heritage study. This may include a search of all relevant cultural heritage registers, a review of relevant published works on the history and ethnography of the area, as well as consultation with the local community in order to determine whether the site is of cultural value to the local people. The significance of any recorded heritage sites in the area should then be assessed, followed by the development of an appropriate management plan.

Figure 3.11 Bushfire protection measures in combination.

Source: NSW Rural Fire Service, Australia

Environmental preservation

Most countries have legislation in place to provide the legal framework to protect and manage the environment on all levels. Parties involved in construction, land clearing, waste management, handling asbestos, etc. have a responsibility under such legislation to assess, report and manage these activities in relation to the surrounding environment and native flora, fauna and ecological communities.

Examples of such legislation frameworks are:

- USA: Environmental Protection Act
- UK: Environmental Protection Act 1990
- Australia: Environmental Protection and Biodiversity Conservation Act 1999 (EPBC)
- India: National Environmental Appraisal and Monitoring Authority 2014
- China: Environmental Protection Administration 1987

The potential impact of solar farms on the surrounding environment will be site-specific and will depend on the scale of the project. It may be a requirement or considered to be prudent to undertake an environmental impact statement as part of any solar farm's due diligence process. Undertaking this assessment may add extra time and cost to the project's preliminary stage and, depending on the level of environmental impact, there may be significant mitigation measures (controlled action) required. Such measures will be assessed and imposed by the relevant government department or minister.

Site assessment and planning

Initial calculations

The site assessment and planning will provide the initial calculations to aid the preliminary design as well as the preliminary estimated costs and returns for the project. It is important for these calculations to be done in the early stages of planning so that finance can be secured early and to avoid unexpected expenses and time delays during commissioning. Some of these calculations may include:

- distance to high-voltage line
- distance to substation
- estimated transmission loss factors
- estimate of the total area required considering the spacing between rows, number of modules, vehicle access, security fence
- initial energy yield using solar resource data and estimates of plant losses based on nominal values from existing projects
- near field and far field shading scene
- cost of land (taxes, fees, rates, etc.)
- access to water and cost of access
- local temperature data to calculate losses
- preliminary geotechnical survey to determine the type of terrain, and hence the type and cost of mounting system required

- wind speed zone to determine wind load to also help determine the type and cost of mounting system required
- logistics cost calculations
- cost of on-site amenities
- cost of lodging for labour
- cost of HSE and security measures
- estimate of equipment costs
- estimate of electrical labour costs
- estimate of civil costs
- estimate of structural costs
- connection fees (DNO or TNO)

Solar resource assessment

When planning a solar farm, it is important to assess the site's solar resource by analysing long-term site-specific solar radiation data sets to understand the spatial and temporal irradiance patterns and how the output of the solar farm will be affected. This is particularly critical for assessing the suitability for the use of tracking applications, as well as for making preliminary yield and performance calculations.

Solar irradiation data

In order to accurately assess the solar resource, it is important to have accurate solar resource data for the site. Global horizontal irradiation is the sum of direct and diffuse radiation and is the most useful parameter for evaluating the solar resource at a given site. Existing long-term data is available from ground-based weather stations and satellite-derived data sets. Some examples include:

- NASA Surface Meteorology and Solar Energy Data Set (for all locations) http://eosweb.larc.nasa.gov/sse
- *Australian Solar Radiation Data Handbook – Exemplary Energy* www.exemplary.com.au/solar_climate_data/ASRDH.php
- Australian Government Bureau of Meteorology www.bom.gov.au/climate/data
- PVGIS EU Joint Research Centre (for Europe and Africa) http://re.jrc.ec.europa.eu/pvgis
- NREL monthly average and annual average daily total solar resource data (for USA) www.nrel.gov/gis/data_solar.html

A world map of long-term average global horizontal irradiation is given in Figure 3.12, illustrating areas of high solar irradiation. Although this kind of map is useful for recognising areas of high irradiation on average, it does not show seasonal, daily or inter-annual variation, which is necessary in order to quantify the uncertainty in revenue from the solar farm in any given year. In order to minimise uncertainty of the long-term solar resource, it is recommended to compare different data sets (i.e. from ground-based measurements or satellite imagery). Ground-based data is typically more accurate but becomes less accurate the further the site is from an existing weather station. Satellite derived data can be used all over the globe and so may be preferred if no ground-based measurements have been taken close by.

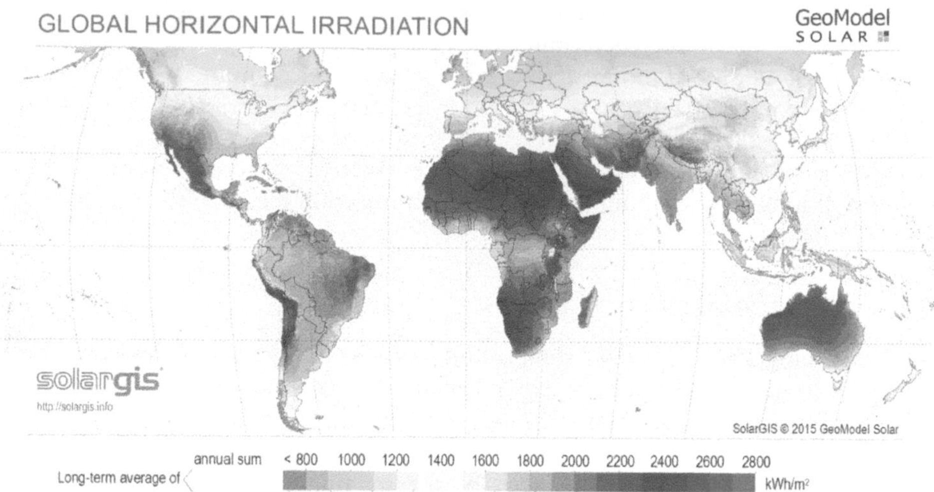

Figure 3.12 This world map of global horizontal irradiation shows which areas in the world have good solar radiation on average. However, it does not indicate seasonal, daily or inter-annual variation, which is very important when sighting solar farms.

Source: SolarGIS © 2015 GeoModel Solar

Figure 3.13 Pyranometer used to measure global horizontal radiation on-site.

Source: USGov (www.nrel.gov)

For the highest accuracy, a land-based sensor such as a thermal pyranometer (see Figure 3.13) should be installed at the site to measure data for 12 months (ensuring seasonal variation is captured). This ground-measured data can then be calibrated against pre-existing data sets to get a more accurate estimation of annual solar radiation.

Measuring the effect of shading

Using on-site solar radiation measurements will account for any reduction in solar access due to natural landscape, such as a valley that cuts out the sun in the early morning and late afternoon.

There are various tools that can be used to conduct a full solar access and shading analysis throughout the year e.g. Solar Pathfinder. This type of tool take 360-degree images of the sky from different points on the site and superimpose them against a latitude-range specific sun path diagram to measure the skyline and any objects that may shade the array at any time of the year. The Solar Pathfinder uses a plastic dome placed over a sun path diagram to reflect a panoramic view of the site. See Figure 3.14 for example outputs for the Solar Pathfinder. Compatible software is available and is used to analyse the results from the Solar Pathfinder and to reproduce this sun path data in tabulated forms for use in the design of the system.

Site boundaries

Site boundaries are important from a legal perspective. There may be building regulations that prevent any structures from being built within a certain distance from the site boundary. Therefore it might be wise to have a survey carried out to confirm the proposed site's boundaries: the survey would mark out the land boundaries and place permanent land markers at the site. The survey document

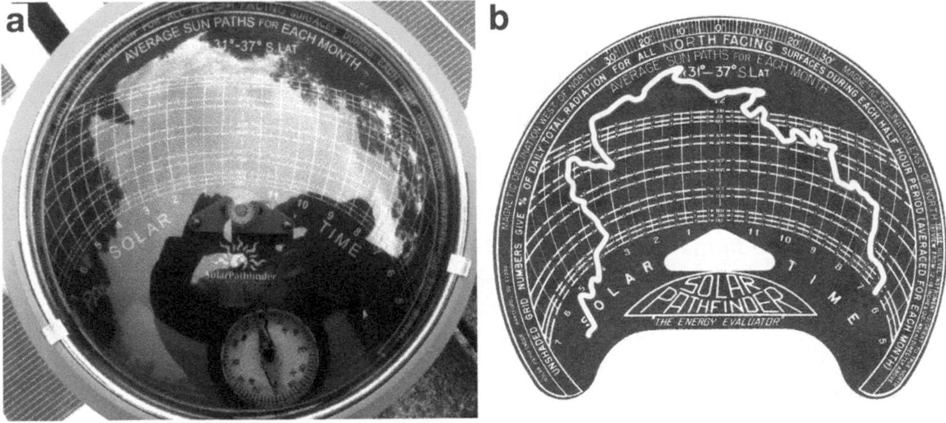

Figure 3.14 (a) Solar Pathfinder showing shading incidence. (b) Sun path diagram showing traced shading line – the area outside the line represents the losses due to shading.

Source: Global Sustainable Energy Solutions

would then be included with the legal documentation forming part of any development application so that the location of the solar farm and associated buildings are accurately interpreted in relation to the overall site. There may also be issues regarding the proximity of the boundary to heritage sites or ecologically sensitive areas, requiring additional mitigation measures in order to gain approval.

The survey would be completed according to a map grid such as the Geocentric Datum of Australia, which is also important for the detailed design stage since the designer will use these permanent markers as references throughout the design.

Appurtenant structures

Appurtenant structures refer to existing structures or utilities on the site that may affect or be affected by the construction of the solar farm. The treatment and impact of these structures must be considered during the early stages of planning. If these structures shade the array or take up valuable space, they may need to be demolished, and so the cost and impact of any demolition must be estimated. Problems may arise if the structure is heritage listed or is of value to the local community. If there are utilities at the site, the construction of the solar farm may be affected, where these are located (overhead, underground, above ground, etc.). For example, if there are underground structures at the site, groundworks may need to be planned in order to work around this existing infrastructure.

Landscape plan

Once the site is selected, a landscape plan can be developed based on the results from various surveys and assessments, including the geotechnical survey and ecological assessment. The landscape plan should include:

- Soil and water management plan to account for land erosion and flooding risks. This is based on the topography and type of soil on the site, as well as the groundwater level and proximity to bodies of water.
- Dust management plan to reduce soiling of the modules from dust.
- Weed management plan to ensure the site does not get overgrown with weeds that may shade the modules or damage native flora.
- Flora and fauna management plan to ensure the protection of local native flora and fauna.
- The landscape plan should also consider the local bushfire protection requirements.

Site access

Adequate site access to the solar farm must be available. Access is required during construction, operation or decommissioning. During the construction period, large vehicles will need to access the site to deliver materials and equipment, so it is necessary to ensure that the local roads are capable of

Figure 3.15 Dunayskaya Solar Park (Odessa Region, Ukraine), shows a well-designed layout, allowing space for maintenance.

Source: Activ Solar

carrying these heavy vehicles and, if necessary, to upgrade roads and intersections or build new ones. New roads should be built with at least a gravel chip finish or equivalent. It may also be necessary to conduct a traffic impact assessment to guarantee traffic safety and accessibility on local roads, and to identify appropriate mitigation measures.

If the site is located on agricultural land it is important to ensure that access tracks throughout the site are kept to a minimum and are only temporary. While some tracks such as those linking inverters may be necessary, vehicles like quad bikes or 4WD should be able to access the site for maintenance without the need to build any tracks.

In order to protect the site against vandalism or damage from large animals, as well as to prevent people from entering for their own safety, a security fence should be constructed around the site boundary. Depending on local regulations, it may be necessary to install a proven security fence that has been tested and approved by government standards. It is important to consider the visual impact of the fence by minimising height and utilising natural features like hedges or crawlers to screen the fences.

Safety and security

All personnel should be aware of all the safety precautions required for working on both construction sites and electric installation sites – as well as the particular hazards associated with installing PV systems. Each site proposed to house a solar farm will have its own set of conditions and installation requirements, and therefore each site will be different. Each component of the system should be

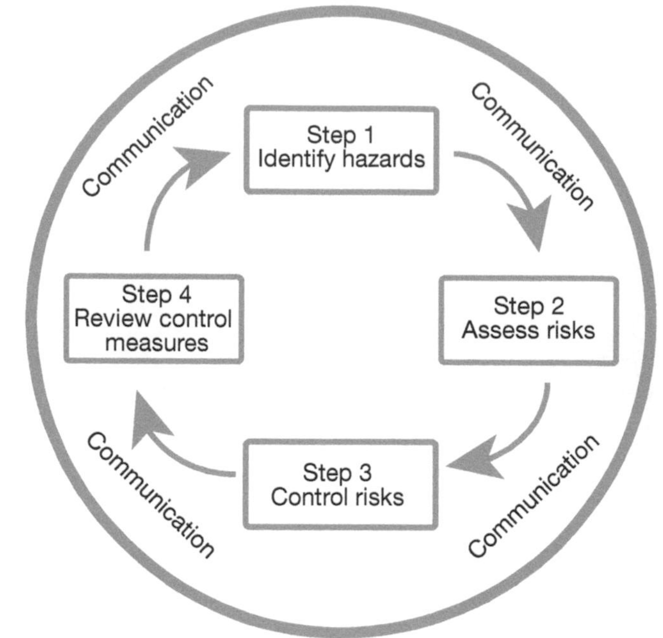

Figure 3.16 The risk-management cycle.

Source: Global Sustainable Energy Solutions

Risk Level

Low | Medium | High | Extreme

Chance \ Impact	Minor	Moderate	Major
Very likely	Acceptable risk	Unacceptable risk	Unacceptable risk
Likely	Acceptable risk	Acceptable risk	Unacceptable risk
Unlikely	Acceptable risk	Acceptable risk	Acceptable risk

Figure 3.17 An example of a risk-assessment matrix. Any risks that are not green must be controlled.

Source: Global Sustainable Energy Solutions

carefully assembled and installed in accordance with the relevant manufacturer's instructions. Because each site is different, safety risks will also vary at each site. Therefore, it is important that detailed health and safety (HS) rules and procedures – also known as work, health and safety (WHS), health and safety executive (HSE) or occupational health and safety (OHS) – are developed which are appropriate to each specific site.

Mitigation of safety risks is a continuous process and is the responsibility of all working on-site. The risk mitigation procedure should use the following steps for each hazard and these should be clearly communicated to all workers on-site:

1 Identify hazards: find out what could cause harm.
2 Assess risks: understand the nature of the harm that could be caused by the hazard, how serious it would be and the likelihood of it happening. A risk-assessment matrix is a common tool for quantifying the severity of the risk.
3 Control risks: implement the most effective control measure that is reasonably practicable in the circumstances.
4 Review control measures: ensure control measures are working as planned.

Physical risks and mitigation methods

Electrical hazards

Site workers on a solar farm are exposed to various electrical risks including AC and DC electricity, different voltage levels, different sources of electricity and different conductive materials. The most significant risk associated with working with electrical equipment is electric shock caused by current flowing through the body which can result in burns or muscle contraction: the higher the voltage, the higher the risk. Muscle contraction is particularly dangerous as it can result in the inability to let go of the live conductor, affecting the heart and sometimes resulting in death. Electric shock can also cause people to fall off ladders or be thrown onto equipment, causing injury. Arcing is another significant electrical hazard which involves electricity 'jumping' through air, smoke or water and is more likely under higher voltages. Arcing may generate substantial heat and ignite flammable substances nearby, which may result in fire, damage to equipment or injury to people on-site.

Some PV system-specific electrical hazards:

- PV array: the PV modules in the array should always be considered live. They generate electricity as light falls on them and attempting just to cover them (e.g. with a blanket) is not a safe practice. Utility-scale PV systems can have an array voltage of up to 1,000 VDC which can be deadly. Different electrical codes specify different maximum voltages. Where possible, ensure that all PV equipment is isolated before commencing work. For example, according to the Australian AS: NZS 5033 2014 if the PV array maximum voltage is greater than 600 V, the cabling to the point of connection to the inverter should be enclosed in heavy duty conduit that cannot be accessed without the use of a tool. People working on or near low-voltage installations must be adequately trained and hold the appropriate licence or licences.

Table 3.2 Typical physical hazards associated with construction and operation of solar farms

Safety hazard	Risk involved	Mitigation method
Heavy objects falling	During installation of large-scale solar, cranes may be necessary to lift heavy equipment, leaving the risk of heavy objects falling onto workers causing serious injury and potentially death	• Ensure hoist has been properly maintained • Ensure load is properly secured • A hard hat should be worn at all times on-site • Workers should never walk under a lift • Operator must always lower a load to the ground before leaving the lift
Sun exposure	PV modules are usually installed in areas with high exposure to sunshine with limited shading. Therefore, there is risk of sunburn, dehydration and sunstroke to site workers during installation and maintenance work	• Wear hats and cover exposed skin • Take regular breaks in the shade • Wear high-factor sunscreen • Drink lots of fluids (preferably water and never alcohol)
Cuts	Many components of a PV system can have sharp edges and cause injury. Some of these components include metal framing, junction boxes, bolts, nuts, guy wires and anchor bolts. Metal slivers from a drill bit can also be hazardous as they often remain around the edges of a hole and can cause severe cuts	• Always wear suitable gloves when handling metal, particularly when drilling
Head protection	Working underneath structures like the array can result in bumps to the head, which can be particularly severe if sharp objects are present. There is also risk of objects falling onto workers if work is being performed overhead	• A hard hat should be worn at all times on-site
Sprains, strains and fractures	Injuries can occur from tripping on rough terrain at the site, particularly when carrying equipment. There is also risk of back strain when lifting and installing very heavy components, such as inverters	• Be aware of site terrain and wear comfortable shoes • Use appropriate lifting techniques including lifting with the legs and not the back • When necessary use equipment to assist with moving heavy equipment
Thermal burns	Metal exposed to the sun can reach temperatures of 80°C which could result in burns if contact is not broken quickly	• Wear gloves • Identify any equipment that should become very hot
Local animal species	Since solar farms are often developed in rural areas, the risk of bites from local species like insects and snakes can be high	• Always be aware when opening enclosures • First-aid treatment for typical injuries from local animals should be known

Source: Global Sustainable Energy Solutions

- Inverter: the output of an inverter is typically in the low voltage range (maximum 1,000 V AC, 1,500 V DC), but is still considered a very dangerous voltage. It is important that all electrical interconnections between the inverter and the switchboard or distribution panel are installed and maintained by a licensed electrician. The inverter should be disconnected from the power transmission path and from the control path for maintenance (if no voltage is required). It is important to wait for the capacitors to discharge completely before opening it.
- Transformer: for a utility-scale plant, a transformer is used on-site to step up the voltage to medium-voltage levels for transmission to the substation,

and then to high voltage for connection to the utility grid. This means there is significant potential for severe injury or damage to equipment at these areas on-site.

- Wiring: connected wiring should always be considered live. Exposed conductor ends should be terminated with tape or cable connectors to prevent anyone from coming in contact with them. If operating equipment such as a crane during installation, the transmission lines on-site should be disconnected in case of metal contact with HV transmission lines.
- Arcing: DC voltages in PV arrays can cause arcing when a contact is opened/broken while the system is live. Arcs are hazardous as they are not self-extinguishing and may cause burns, flashover and electric shock. It is very important that disconnection procedures are carefully followed, and that all switches, disconnects and isolators are rated for load breaking to reduce the chance of arcing occurring.

General electrical safety practices include:

- Wearing suitable personal protective equipment for all electrical works for shock protection including appropriately insulated gloves and dielectric boots.
- Ensuring that no unauthorised person has access to the enclosed electrical operating area.
- Using suitable tools when working on electrical equipment.
- Building the PV power plant as a closed electrical operating area.
- Switching the PV modules to insulated operation before entering the PV field.
- Installing appropriate warning labels on all electrical equipment.

Electrical safety guidelines and regulations

- Occupational Safety and Health Administration (OSHA) regulations – USA
  - Title 29, part 1910, subpart S – General Industry, Electrical
  - Title 29, part 1926, subpart K – Construction Industry, Electrical
- NFPA 70E: Standard for Electrical Safety in the Workplace – USA
- Electricity at Work Regulations 1989 – UK
- Safe Work Australia Managing Electrical Risks in the Workplace Code of Practice – AUS

Finance and economics

Economic trends

Up until recently, the costs of equipment, construction and system operation have hindered development of the solar PV industry. Since 2015, new materials and technology developments have resulted in efficiency improvements and reduced technology costs, with the price of modules decreasing by 65–70%

between 2009 and 2013. This price reduction has improved the cost competitiveness of solar PV, allowing significant increase in market penetration. Looking at Figure 3.18, costs are expected to continue decreasing up to 2020, predominantly due to reductions in module costs as well as balance of system (BoS) costs. Since the capital cost of solar farms exceeds the operating costs, this reduction in technology costs has allowed a rapid fall in the levelised cost of electricity (LCOE), now ranging between USD 0.15–0.35/kWh for utility-scale projects.

These cost reductions have reduced the risk profile of utility-scale solar installations and have led to significant increases in investment in solar PV technology over recent years. Investment boomed in 2011, but has steadily dropped in the years that followed until 2014, predominantly due to reductions in technology costs and political uncertainty in markets like the US and the UK. However, investment in terms of kWp has continued to increase since 2012. Although China contributed to over one-quarter of the world's share in renewable investment in 2014, another major reason for this growth was the spread of renewable energy to emerging markets in developing countries. Some of these emerging markets include those in Indonesia, Chile, Kenya, Mexico, South Africa and Turkey, all of which invested at least USD 1 billion in renewable energy technologies in 2014. Investment in solar increased by 25% to USD 150 billion compared to 2013, accounting for a record breaking 46 GW of Solar PV capacity installed worldwide. Trends in investment of renewable energy by region are illustrated in Figure 3.19.

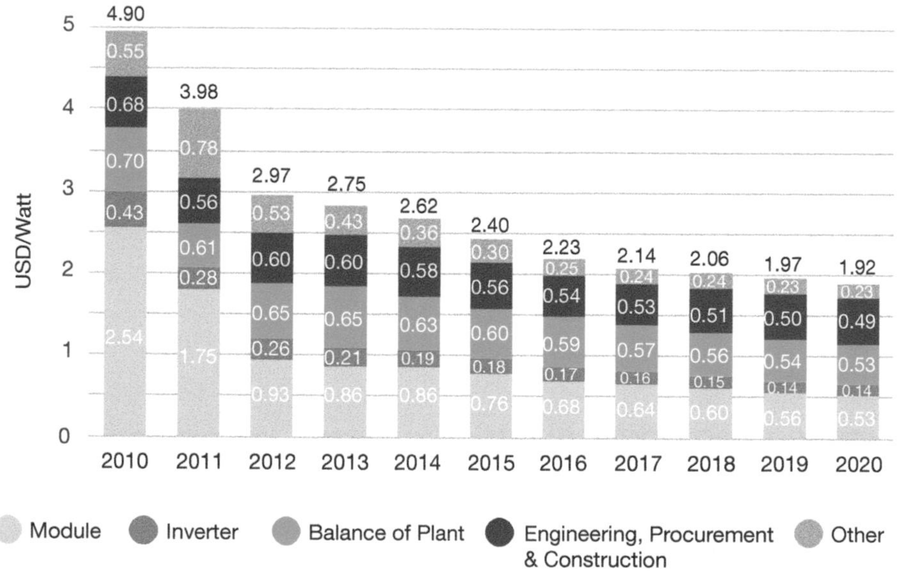

Figure 3.18 Projected utility-scale solar PV system deployment cost (2010–20).

Source: IRENA

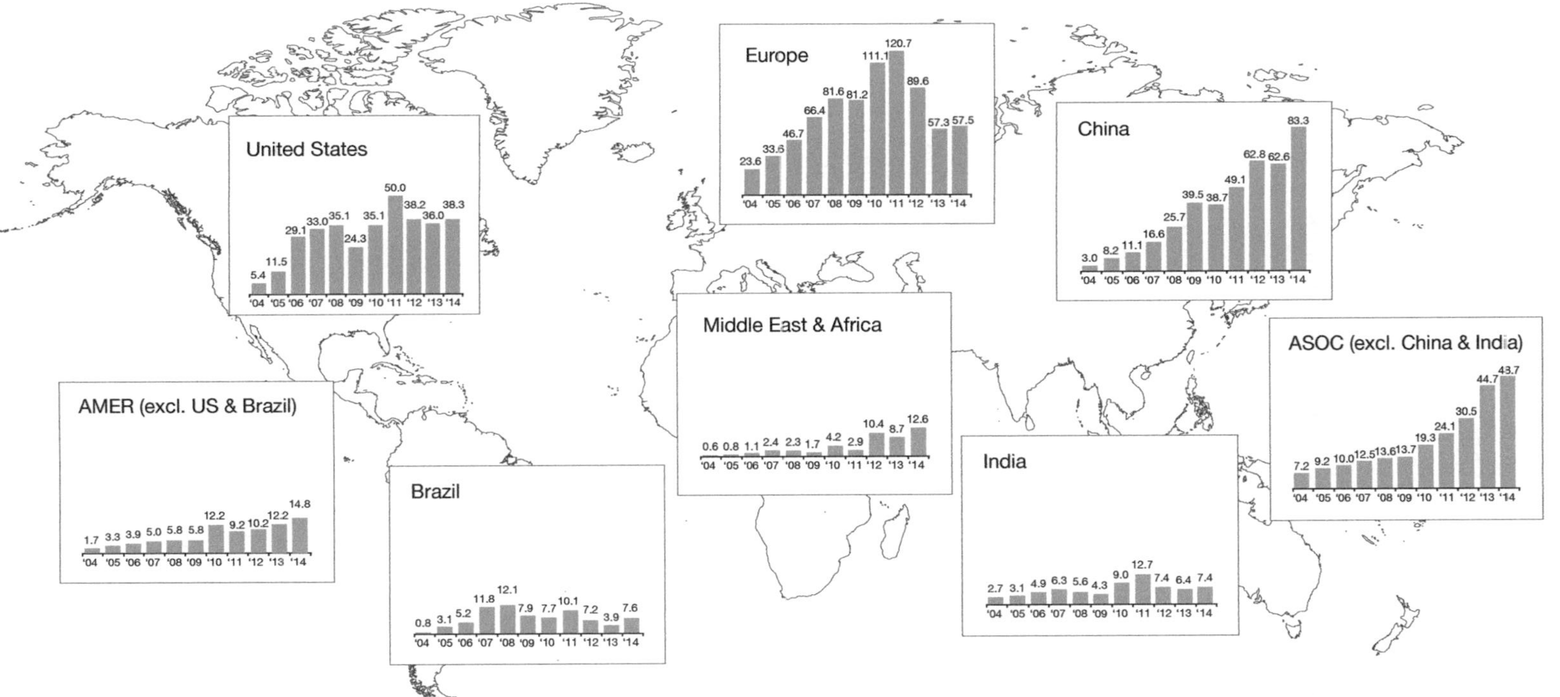

Figure 3.19 Global new investment in renewable energy by region (2004–14) (USD bn).

Source: Frankfurt School-UNEP Centre, 2015

Project financing

Investment risks in renewable technologies have decreased significantly over recent years. This is because of reduced capital costs, increasing experience with the technology, and government support. Future investment will depend largely on whether the market conditions are attractive to private investors such as developers, commercial banks and institutional investors, who currently make up the bulk of finance for solar farms. However, as PV markets and technology continues to improve, it is becoming attractive to a wider range of investors including private equity firms and governments.

Project financing for large-scale solar typically comes from a combination of equity and debt. Equity contributions come from institutional investors, private equity firms or venture capitalists, which contribute capital and are entitled to a share of the income generated from the solar farm. Debt finance comes in the form of loans, typically from commercial banks, and needs to be repaid throughout the project lifetime. It is very important for the developer to engage early with financiers, particularly where there isn't familiarity with solar projects, as these negotiations can take much longer.

In preparation for the due diligence process, the developer will need to prepare comprehensive project documentation. While an economic analysis will no doubt form part of the financier's due diligence process, the developer should also perform this analysis in order to ensure the project is feasible and to gain initial interest from financers. This will include the calculation of important metrics such as the levelised cost of electricity (LCOE) and internal rate of return (IRR).

LCOE is a metric used to evaluate the life cycle costs of a particular type of generation technology or project per kWh or MWh of electricity generated throughout the project lifetime. The calculation includes the capital costs, O&M costs and fuel costs. It is useful to calculate and estimate the LCOE during the planning stage to help secure finance.

LCOE can be calculated using the following formula:

$$\mathrm{LCOE} = \frac{\sum_{t=1}^{n} \frac{I_t + M_t + F_t}{(1+r)^t}}{\sum_{t=1}^{n} \frac{E_t}{(1+r)^t}}$$

where

I_t = Investment expenditure in year t
M_t = Operations and maintenance expenditures in year t
F_t = Fuel expenditures in year t
E_t = Electricity generation in year t
r = Discount rate
n = Life of system (amortisation period)

IRR is a metric used to calculate the financial benefit of an investment. It is the equivalent interest (discount rate) that would allow the project to break even.

This is calculated by working out the interest rate that will make the net present day value of both the project costs and income equal to zero. The higher the IRR, the more attractive the investment becomes. It is a complex calculation, and should be calculated using a tool, such as the IRR function in excel.

Due diligence

Before the project finance agreements are finalised, the finance partners will conduct an evaluation of the project (due diligence) covering the legal aspects, permits, contracts (EPC and O&M), and specific technical issues, in order to determine the risks and mitigation methods prior to investment. More details on the different types of risks that may be assessed during due diligence are given later in the section on project risks.

There are typically three main due diligence evaluations:

- Legal due diligence – assessing the permits and contracts (EPC and O&M).
- Insurance due diligence – assessing the adequacy of the insurance policies and gaps in cover.
- Technical due diligence – assessing technical aspects of the permits and contracts.

These include:

- ○ sizing of the PV plant;
- ○ layout of the PV modules, mounting and/or trackers, and inverters;
- ○ electrical design layout and sizing;
- ○ technology review of major components (modules/inverters/mounting or trackers);
- ○ energy yield assessments;
- ○ contract assessments (EPC, O&M, grid connection, power purchase and [FiT] regulations); and
- ○ financial model assumptions.

However, due diligence may only take place once the following tasks have been completed:

- site selection;
- negotiation of land use;
- initial solar resource assessment;
- environmental impact assessment;
- initial layout and design including initial equipment selection;
- planning permits;
- grid connection offer or letters of intent; and
- application for FiT and/or PPA.

The due diligence process requires that a lot of preliminary work is completed before the project can receive any confirmation of finance. Once due diligence is completed, the developer will have to work to ensure that the final design adheres to the requirements of the lenders, and that the plant operates at the performance specified in the agreement.

Finance options

Finance options for solar farms vary across different countries depending on tax structures, policy and levels of government support. In some countries, institutions have been set up to help reduce financial barriers to developing solar farms, while others are set up to take advantage of government incentives or expand access to a wider range of investors.

Examples:

- Australia
  Government organisations like ARENA and the CEFC are reducing some of the financial barriers to support the uptake in utility-scale solar across the country. The CEFC helps projects overcome financing hurdles by providing finance where gaps are present and to open up the space for other financiers.
- United States
  Finance for solar farms in the US is designed to maximise the value of tax benefits offered by state and federal governments. This is known as tax equity investment and may be achieved through partnership flips which may be 'all equity' or 'leveraged' with debt; or through lease structures. A 'partnership flip' is a financing arrangement between an investor and a renewable energy developer. This mechanism is popular for financing projects to maximise the government-based tax incentives to improve the economics of these projects. The aim of these financial structures is not only to make good use of capital and government incentives, but also to shift the project uncertainties to the party most willing and best suited to take the risk. A typical financial structure is illustrated in Figure 3.20, where the green boxes represent distribution of risk to parties involved in the project development and operation.

Yieldcos have emerged as a popular alternative to the typical finance structure in the US. A 'yieldco' is a publicly traded company that owns a group of renewable (or other) operating assets that generate predictable cash flows to raise capital. This cash can then be distributed to investors as quarterly or yearly dividends. The capital raised from the smaller operating projects can be used to finance new projects at lower rates than through tax equity finance. This structure is considered low risk because cash flow comes from assets that are already operating and so the projects can attract a wider range of investors. However, there is uncertainty in future cash flow or access to assets and lower capital typically means lower returns on investment.

There is a range of financing options for solar farms, but each one requires thorough research, investigation and marketing to determine and secure the best options for a particular project. It is important to take advantage of government incentives and institutions set up to increase market penetration of utility-scale solar.

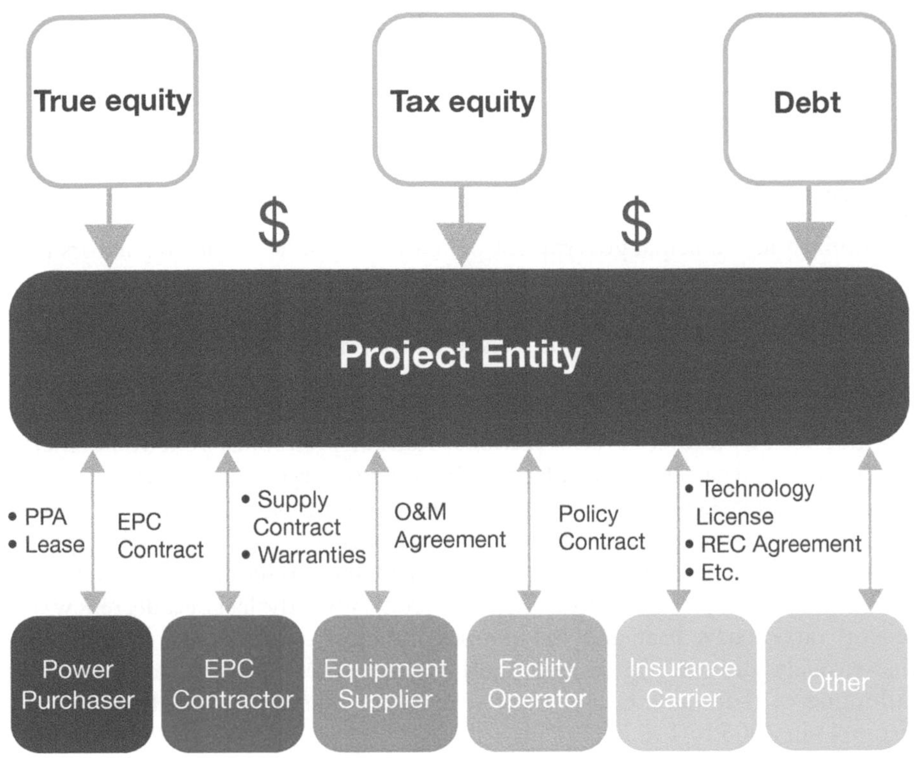

Figure 3.20 Typical PV financial structure in the USA.

Source: NREL

Selling electricity

Assurance for the sale of electricity is naturally the most critical aspect of a successful solar farm, so it is important to consider all of the options available to ensure that debt will be paid off and that the solar farm is profitable. Generally, the solar farm will not attract finance until it has secured the method by which the electricity will be sold.

Most commonly, electricity generated from a solar farm is sold via a power purchase agreement (PPA), which is a contractual agreement between the developer and a utility (or other buyer) for the sale of electricity at a fixed price over an agreed period (for more information see the section on power purchase agreements).

Another option is to sell the output in the wholesale electricity market, through which the utility or commercial buyer, depending on the market model available, purchases the electricity at the wholesale market price (USD/MWh). This price can be set by the market to be the cost of generation of the most expensive generator operating during a given period of time in order to meet demand. There are various market models employed internationally, one being that the cost of generation is set by the short run marginal costs (SRMC) of the generator, which are essentially the operational costs of the power plant. If the SRMC of a generator is below the wholesale electricity price during a given time period, it can dispatch electricity to the grid. Because solar farms do not incur any fuel costs, the operational costs and hence the SRMC are very low, therefore when operating, they can almost always sell electricity in the wholesale market. While this means that the solar plant is subject to fluctuations in wholesale electricity market prices, and hence uncertain revenue, this can be a good alternative when PPAs prove difficult to negotiate.

Government policies and their financial impact

National and municipal government policies and public service regulators play a vital role in creating a supportive environment for investment in renewable energy. Effective support policies create a market for renewables by reducing investment risks and enabling utilities to adapt business models which facilitate the integration of renewable energy technologies.

Government-based incentive policies to support investment in solar PV include feed-in tariffs, reverse auctions, market premiums, renewable energy certificates and tax incentives. Renewable energy targets play a crucial role in quantifying market potential and therefore facilitating investment in utility-scale solar. However, any of these mechanisms must be backed by specific policies and support measures in order to provide a reliable trajectory for the technology deployment for investors and society. Over the last decade, renewable energy targets have been adopted in more and more countries and have become increasingly diverse. Figure 3.21 and Figure 3.22 illustrate the growth of adoption of renewable energy targets, with the number of countries growing from 43 in 2005 to 146 by 2015.

Feed-in tariffs

The most popular incentive scheme to date has been feed-in tariffs (FiTs) that provide fixed revenue to solar farm developers per MWh of electricity generated. FiTs have been used to meet renewable energy targets (RETs), by providing long-term policy stability to attract investors, helping to create jobs and foster economic development. If FiT policies are well designed, they can prove to be an economic solution to encourage development of RE power plants such as solar farms, providing benefits to taxpayers, developers and society in general.

FiTs are typically paid at a fixed rate over a defined period (15–20 years), and may be calculated in one of the following ways:

1 Based on the actual levelised cost of generation plus a target rate of return, ensuring the project is profitable.
2 Based on the utility's avoided costs, i.e. the estimated costs of electricity if it were supplied by a different generator.
3 Fixed-price incentives that are awarded arbitrarily, i.e. not based on LCOE or avoided costs.

The first is the most common method used around the world, and has proved particularly successful across Europe. The second method has been more commonly used in the US, but has shown limited success, as it does not necessarily provide investors with a price that guarantees adequate return on investment. With the price of solar technology falling rapidly, FiTs should be

Figure 3.21 Global map of national renewable energy targets of all types (2005).

Source: IRENA, 2015

Figure 3.22 Global map of national renewable energy targets of all types (2015).

Source: IRENA, 2015

designed to decrease accordingly over time, which can help stable market growth, enhance investor confidence and help solar reach grid parity.

Challenges surrounding FiT policies include:

- failure to assist in covering high capital costs of solar farms;
- potential to increase electricity prices;
- significant administrative procedures to design a suitable FiT policy (particularly if based on LCOE or avoided costs);
- the possibility of poor project siting if projects are guaranteed a connection to the grid regardless of location;
- requirement of periodic adjustment due to rapid changes in technology costs and market prices;
- projects are very reliant on government incentives.

Reverse auctions

In recent years, there has been a shift away from FiTs to competitive auctions or tenders. These alternative processes have been driven largely by the desire to reduce reliance on government subsidies and allow for market-driven incentives. Reverse auctions use the auction's bidding process, but the roles of seller and buyer are reversed: the sellers bid for the prices at which they are willing to sell their goods and services. A reverse auction for utility-scale solar projects would see governments issue tenders or request bids for the acquisition of projects from which a utility or network service provider will purchase electricity. The reverse auction will typically be designed for the acquisition of a certain capacity of renewable energy projects, and the government will continue to award winning bids until the desired capacity has been met.

Germany, a global leader in renewable energy, has recently introduced a policy shift from FiTs to reverse auctions with the hope to utilise brown land and unproductive agricultural sites, as well as drive down PV prices in that country. The new German policy is a pilot project running through to 2017, starting in 2015 with three auctions making up 500 MW of installed capacity. Auctions will typically be won by the bidder willing to sell electricity generated from the solar farm for the lowest price, provided all requirements for the project are met. Bids are not to exceed €0.1129/kWh, with the aim of reducing the cost of solar energy. The shift has not been welcomed by all, according to German Wind Energy Association BWE: 'All of the experience abroad shows that the German government's three goals – lower costs, a diversity of market players, and target attainment – cannot be reached in reverse auctions'. If auctions end up producing higher prices, it is unlikely that auctions will continue to replace FiTs entirely. Auctions are not unique to Germany and have been introduced in many other countries including Brazil, Peru, South Africa, China, Australia and India, as well as other parts of Europe.

Reverse auctions can reduce overall policy costs by increasing competition and procurement efficiency, reducing government financial assistance, and by more accurately reflecting prices in rapidly changing markets. However, any success will depend on how the policy is designed and the market conditions for the countries and jurisdictions in which the policy is deployed.

Market premiums

A market premium (also known as a contract for difference or a premium FiT, depending on which country) is a hybrid policy instrument aiming to provide both investment security and wholesale integration. A market premium is a payment that may be awarded to renewable energy generators that sell electricity in the wholesale market, to provide more revenue security than the wholesale market offers. Market premiums have been designed in Europe to help reach ambitious renewable energy targets in a way that minimises the cost to taxpayers, while maintaining market competition.

Spain was the first country to implement a form of market premium, known as a premium FiT, and this was originally introduced as a fixed payment on top of the wholesale market price. However, this measure led to overcompensation for RE generators during periods of high wholesale electricity prices, and high risk for generators during periods of low wholesale prices. For this reason, policy makers in Spain have introduced cap and floor prices, which are upper and lower limits on the price that generators can sell electricity.

A contract for difference (CfD) is another type of market premium, which guarantees RE generators a monthly premium payment that makes up the difference between a fixed 'strike price' and the average wholesale market price over the course of that month. See Figure 3.23 for an illustration of how this arrangement is planned to operate in the UK. A similar approach has been adopted in Germany.

So far, market premiums have been successful in increasing the volume of renewable energy generation in the wholesale market.

Renewable energy certificates (RECs)

Tradable renewable energy certificates (RECs) or tradable green certificates (TGCs) are digital or hard copy documents that represent the non-power attributes

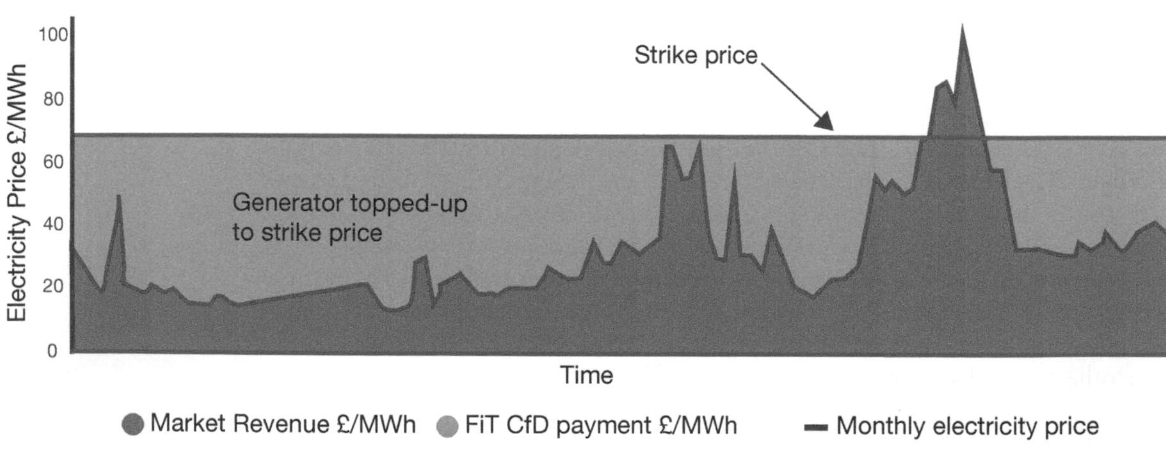

Figure 3.23 How the CfD is planned to function in the UK.

Source: Clean Energy Solutions Centre

of renewable energy generation such as environmental and social benefits. RECs are typically issued by a country or states' energy regulator and may be bought, traded or forfeited by electricity utilities and retailers. RECs provide a revenue stream for renewable energy generators in addition to the sale of electricity to improve the financial viability of renewable projects. RECs are a common policy mechanism used to monitor and help reach renewable energy targets.

In the US, one REC represents 1 MWh of electricity generated from a renewable resource, and can be used to monitor compliance with a state's renewable energy target established under the Renewable Energy Standard. Utilities or electricity suppliers can use RECs to prove that they have procured or generated a certain amount of renewable energy to comply with the mandate.

In Australia, large-scale generator certificates (LGCs) are a form of RECs, and work in a similar way, representing 1 MWh of renewable energy generation, awarded to generators. Retailers are required to purchase a certain number of RECs in order to help meet the national renewable energy target. Given that the maximum LCOE of utility-scale solar is significantly higher than the wholesale price of electricity in the Australian National Electricity Market, RECs are imperative to helping these projects become viable. According to BREE forecasts (see Figure 3.24), even with the sale of RECs, even the most efficient solar projects will not be viable until 2018.

Tax credits

Tax incentives have been implemented in over 25 countries to encourage the growth of renewable energy in the generation mix. Investment tax credits (ITCs) reduce the amount of tax based on capital expenditure for which solar farm

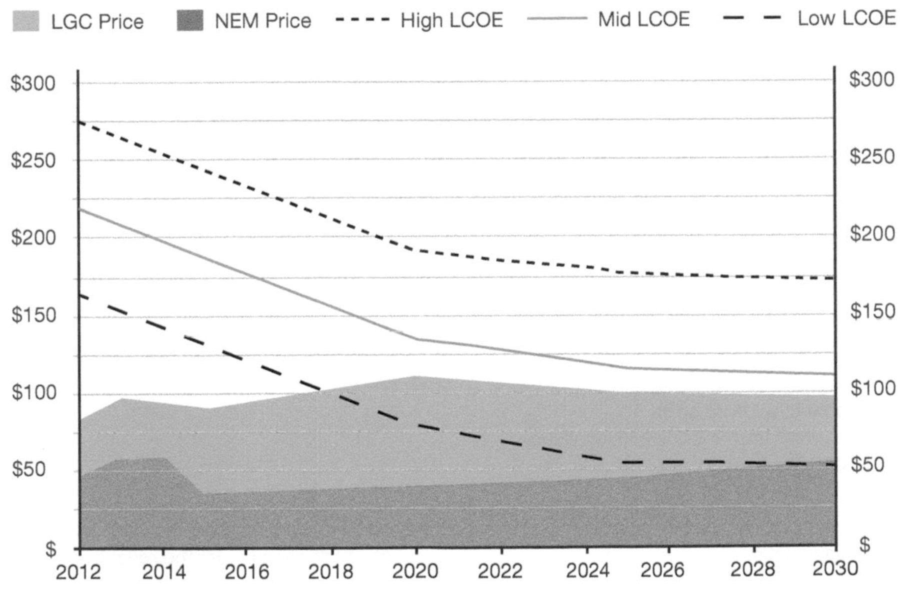

Figure 3.24 Comparison between forecasted levelised cost of solar PV and available revenue.

Source: Bureau of Resource and Energy Economics, ACIL Allen

owners or developers are liable. ITCs are therefore very effective in addressing some of the risks associated with financing renewable energy generators, which typically have a high capital cost, particularly those in the early stages of deployment. On the other hand, production tax credits (PTCs) are awarded throughout the project lifetime based on the amount of electricity generated. This means that PTCs reward optimal performance, so that developers choose to use high-quality equipment and ensure that adequate and regular maintenance is upheld throughout the lifetime of the project.

Because tax incentives reduce government tax revenue, policymakers may introduce a maximum incentive to cap the volume of credits awarded. This will depend a lot on a specific country's circumstances. The tax incentive period can also be adjusted to suit the goals of different jurisdictions. For example, a short timeframe can incentivise rapid deployment; whereas a long period will incentivise long-term project developers.

Environmental conservation

Environmental issues associated with solar PV projects and risk management

Although solar farms can be seen as having a positive environmental impact by replacing fossil-fuel generators with cleaner renewable generators in the electricity mix, there can also be some adverse impacts associated with the construction, operation and decommissioning of solar farms. Some of these impacts and associated mitigation measures are set out in Table 3.3.

Table 3.3 Potential environmental impacts and possible mitigation measures

Environmental issue	Description of impact	Mitigation measure
Vegetation clearing	• Loss of habitat for native flora and fauna • Potential to spread weeds • Increase chance of soil erosion	• Clearing vegetation should be avoided as much as possible • Vegetation and habitat restoration plan should be put in place after construction is completed
Soils and landfills	• Soil compaction • Erosion from land clearing • Erosion from module runoff in concentrated area	• Install silt fences • Construct rock/straw barriers to prevent runoff from entering existing bodies of water • applying erosion blankets to control flow
Surface runoff and flooding	• Disturbance of natural drainage channels • Contamination from spills • Water runoff from modules in localised area can channelise stormwater flows and increase the velocity	• Properly manage excavation to allow stormwater to drain properly
Glint and glare	• Can cause discomfort to viewers • Can impact aircraft safety	• It may be necessary to complete a glint and glare assessment to consider the reflectivity of the modules, frames and supports

Noise and ground vibration	• Disturbance to local residents during construction	• Avoid vehicle and equipment idling • Construction activities should only take place during normal business hours • Equipment will be in good condition with use of muffling devices where applicable
Visual impact	• Loss of landscape amenity • Impact on scenic character of the area for residents and motorists	• Retain existing vegetation like mature trees and hedges when possible • Screen fences or buildings that house electrical equipment with vegetation • Design so that modules follow contours of landscape

Source: Adapted from NGH Environmental

Environmental impact analysis

An EIA is an assessment of all positive and negative impacts that a project may have on the environment for the project's lifetime, and is generally required for all large projects. It should be completed by a qualified environmental impact assessor and if any significant impacts are identified, approval may be required by a government environment minister.

An EIA should address all impacts described in the previous section, and assess:

1 The magnitude and severity of each impact.
2 The value of the resource or organism that will be impacted.
3 The duration and reversibility of the impact.

The environmental assessment process under the Australian EPBC is illustrated in Figure 3.25. Other countries will have similar procedures.

Project risks

There is a variety of project risks that exist throughout the life cycle of a solar farm that affect and can be attributed to a number of parties including developers, investors, manufacturers, site workers etc. It is critical to identify all project risks at the early stages of development so that they can be mitigated and allocated to responsible parties.

All projects will need adequate insurance including general liability and property insurance. Additional insurance can also be taken out against other risks if necessary, or required by project financiers. This may include performance insurance, environmental risk insurance and business interruption insurance. However, insurance is typically a very expensive method of risk mitigation. A study by NREL in 2010 found that insurance premiums can make up about 25% of a PV system's annual operating costs. It is very important to weigh up the benefit of a specific type of insurance against the price. Note that this study

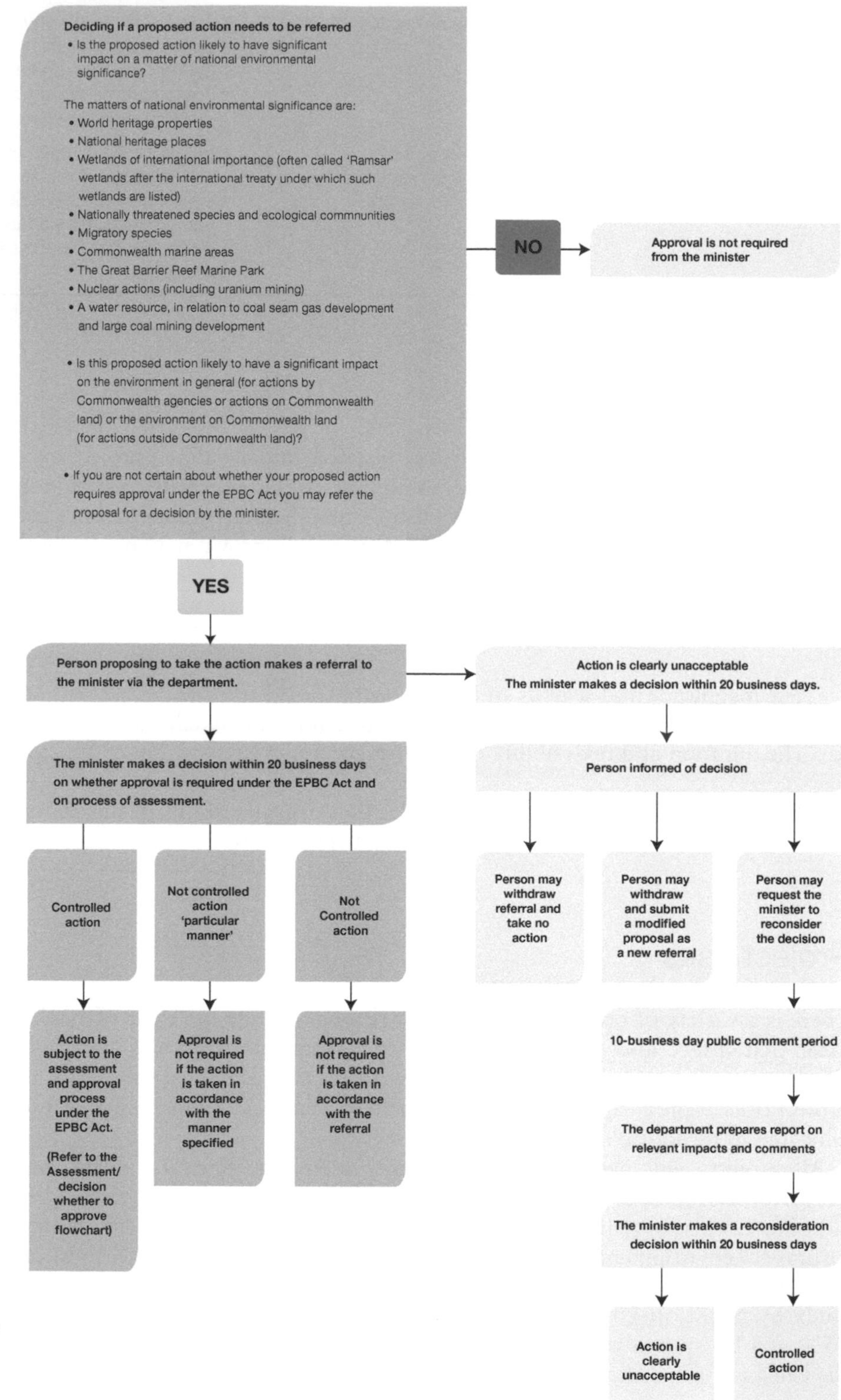

Figure 3.25 EPBC Act environment assessment process – referral.

Source: Australia Government, Department of Sustainability, Environment, Water, Population and Communities

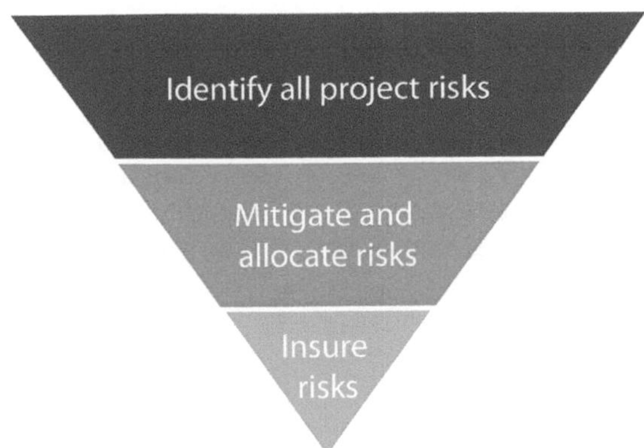

Figure 3.26 General risk management strategy for solar farms.

Source: NREL

was completed in 2010, and it is likely that insurance premiums have decreased as insurers have become more familiar with PV projects.

It is important to note that the level of uncertainty or potential damage from an individual risk won't necessarily disallow the project from gaining finance, as long as other risks are managed effectively. The project risks and mitigation measures will be assessed during due diligence and will influence the project financing options.

Project risks can be broken up into risks that occur during the development stage and those that occur during operation, each with technical and non-technical counterparts. The project risks outlined in this section cover the main considerations and should be used as a guideline only. Each project will have different risks depending on the location, incentive regime, developer, capital structure, etc.

Development risks

Development risks are those that occur during the planning and construction phase of the project. Technical development risks may include those associated with system design, resource estimation, site evaluation and grid interconnection. An outline of these risks and the potential damages associated with those risks is given in Table 3.4.

The most significant non-technical development risks are macroeconomic risks and policy/regulatory risks. These risks are assessed during due diligence because they affect how certain investors can be that they will receive their expected returns. Insurance to protect against policy uncertainty is very hard to come by, and if available, very expensive. All the risks given in Table 3.5 could affect a financier's decision to lend or invest in the project and hence determine the project's availability of capital.

Operational risks

Operational risks are those risks that occur once the project has been commissioned through to the end of the project life. Since solar energy is an intermittent, variable resource, operational risks revolve around the uncertainty

Table 3.4 Technical risks during project development and associated potential damages or losses

Risk	Considerations	Potential damages or losses
Resource estimation	• What level of confidence should be applied to historical solar data?	• Resource-related production shortfalls • Debt service delinquency or default
Component specifications	• What is the performance history or specification of the selected product?	• Manufacturer insolvency • Serial defects
System design	• How well is the system design integrated with the components? • Does it ensure reliability, availability and maintainability?	• Component failures • Production shortfalls • Forced downtime
Performance estimates and acceptance/ commissioning testing	• How well validated are the performance estimates? • What tests are done to confirm baseline performance?	• Production shortfalls relative to estimates • Not enough revenue generated to cover loan repayments
Site characterisation	• What is known and what might not be known about the site? • What are the weather, water, geotechnical and infrastructure conditions?	• Environmental constraints/prohibitions • Infrastructure constraints • Transmission cost overruns
Transport/installation risks	• Are components shipped and installed according to best practices?	• Equipment damage • Project delays

Source: Adapted from and with permission of the National Renewable Energy Laboratory, from www.nrel.gov/docs/fy13osti/57143.pdf (accessed 4 March 2016)

Table 3.5 Non-technical risks during project development and associated potential damages or losses

Risk	Considerations	Potential damages or losses
Transmission/distribution and interconnection	• Is it available? • At what cost?	• Cost overruns • Project delays
Developer risk	• Do they have experience with the technology, project size, and type? • How strong is their balance sheet?	• Developer insolvency • Cost overruns • Project delays • Project abandonment
Power purchase agreement (PPA) and pricing	• Does the project have guaranteed revenues? • Will the project still be profitable under negotiated price?	• Underbidding • Power off-taker insolvency • Project does not secure PPA
Performance estimates and acceptance/ commissioning testing	• What are the possible losses/damages and their associated costs during the construction phase? • What about damage to equipment during transport? • Is the engineering, procurement, and construction (EPC) firm experienced and does it have a strong balance sheet? • Will the project be completed on time and within budget? • Are all the necessary permits in place?	• Injuries • Property/equipment damage • Fire • Weather and natural disasters • Cost overruns • Project delays • EPC insolvency

Policy/regulatory risks	• What are the local regulations for the project and its contracts? • How difficult is the permitting process? • What type of government incentives are available and are they stable?	• Regulator rules against contracts • Incentives change/expire • Project delays • Cost overruns • Failure to access incentives • Failure to obtain permits
Insurability	• Can the developer access insurance? • If so, at what cost?	• Failure to obtain insurance • Insurance too expensive • Uncovered risks
Site control	• Does the developer own the development location?	• Failure to obtain site control and therefore financing, incentives, and other benefits • Project delays • Cost overruns
Multi-contracting risk	• Are risks clearly allocated to the various project parties through contracts?	• Uncovered losses • Lawsuits
Commodities risk	• Are the commodities necessary for project construction (e.g. steel, silicon) available at sustainable cost and are their supply chains secure?	• Price volatility • Unavailability of necessary materials

Source: Adapted from and with permission of the National Renewable Energy Laboratory, from www.nrel.gov/docs/fy13osti/57143.pdf (accessed 4 March 2016)

that the plant will generate enough electricity to earn enough revenue to cover loan repayments. The performance of a solar farm can be quantified by the performance ratio (PR) which indicates the quality and reliability of the plant irrespective of location and solar insolation. The PR of a PV system is expressed as a percentage or a fraction of the theoretical possible output of the PV plant. It takes all the losses in the plant into account. It indicates the actual output of the array compared to the maximum theoretical output.

For example, if the PR of a 100 MWp PV plant is 70% (or 0.7) and the average solar irradiation at the site is 4.5 PSH, then the annual output of the plant is given by:

Maximum theoretical annual output = 100 MWp × 4.5 PSH × 365 days = 164,250 MWh

Actual annual output = 164,250 MWh × 0.7 = 114,975 MWh

A high-performance solar farm can have a PR of up to 0.8.

The developer is required to guarantee a certain PR throughout the lifetime of the solar farm to the investors, and may be required to compensate if the system performance falls below the guaranteed PR. The EPC contractor will often guarantee a PR and will ensure maintenance is performed so that the PR is met throughout the lifetime of the plant, but this may not always be sufficient to cover the developer's performance guarantee obligations. The module manufacturer will also provide a degradation warrantee and will be responsible for equipment repair but will not be liable for liquidated damages. See Table 3.6 for a summary of the technical operational risks, considerations and potential damages.

Table 3.6 Technical risks during project operation and associated potential damages or losses

Risk	Considerations	Potential damages or losses
Operations and maintenance (O&M) risks	• What are the component failure/reliability risks? • Is there adequate availability of spare parts in inventory or rapidly available? • What is the strength of the reliability assessment? • What is the strength of the system design and production engineering? • What is the production availability? • Are there equipment warranties? Are the manufacturers still able to service them? • How strong is the O&M provider?	• Serial failures • Latent defects • High rates of degradation • Module delamination • Forced outages • Planned and unplanned maintenance downtime and costs • Manufacturer insolvency • Resource inadequacy
Power purchaser infrastructure risks	• Is the power purchaser adequately equipped to integrate solar power resources?	• Power export curtailment • Inability of grid operator to handle variability

Source: Adapted from and with permission of the National Renewable Energy Laboratory, from www.nrel.gov/docs/fy13osti/57143.pdf (accessed 14 March 2016).

Table 3.7 Non-technical risks during project operation and associated potential damages or losses

Risk	Considerations	Potential damages or losses
Credit/default risk	• Do production shortfalls threaten the project's debt service? • Is the sponsor's balance sheet strong enough to cover these shortfalls?	• Developer/sponsor default • Developer/sponsor insolvency
Power purchase price risk	• If there is a PPA, is the PPA price high enough to address project costs? • If operating in the electricity market, are regional power prices going up or down, or are they stable? • Are PV installed costs trending down?	• Price makes completed project uneconomic • Market prices for PV systems decline, making project uneconomic in the future • Regional power prices decline, making PV project uneconomic
Power purchaser risk	• What is the creditworthiness of the power purchaser? • Will they uphold their end of the contract?	• Power off-taker insolvency • Power export curtailment
Duration of revenue support	• Debt amortization period should be no longer than PPA period minus two years	• Production shortfalls relative to estimates • Not enough revenue generated to cover loan repayments
Insurance	• When insurance must be periodically renewed, how do cost uncertainty and future price increases affect project economics? • Is the insurance required by the project documents commercially available? • Does the policyholder fully understand terms of coverage? Can they navigate the claims-making process and will insurance company pay out?	• Uninsured losses or damages • Unsuccessful claim • Changes in policy costs make project uneconomic
Transport/ installation risks	• Are components shipped and installed according to best practices?	• Equipment damage • Project delays

Source: Adapted from and with permission of the National Renewable Energy Laboratory, from www.nrel.gov/docs/fy13osti/57143.pdf (accessed 14 March 2016)

Non-technical operational risks reflect the project's ability to sell electricity and pay back loans/investors throughout the project lifetime. These risks also have a big impact on a financier's decision to invest in or loan money for the project. Some examples of these risks are given in Table 3.7.

Solar farm project: application and contractual aspects

Request for proposals and reverse auction

A request for proposal (RFP) is an invitation for professional, qualified solar farm developers to submit proposals for the financing, design, engineering, construction, operation and maintenance of a proposed solar farm. The RFP or reverse auction (as described earlier) may be issued by government bodies in order to reach specified renewable energy targets or by utilities that are mandated to procure a certain volume or capacity of renewable energy generation to reach these targets. Under this tendering process, respondents submit bids to win the right to a FiT from the government, or a PPA with a utility or similar.

Application submission

The proposal should be completed so that it adheres to the requirements and instructions set out in the RFP. It may consist of but is not limited to the following components:

1 Transmittal letter
   - Summary of ability and desire to complete the project
   - Summary of ability to meet the required qualifications, standards and regulations
   - Summary of project milestones including dates and how each will be met
   - Proposal period
   - System description and summary
      - Total system size
      - Number of panels and inverters
      - Panel wattage and inverter size
      - Annual solar panel degradation factor
      - Major component manufacturers
      - Total estimated cost
      - Annual estimated output for each contract year
      - Total projected annual contract costs
      - Bidder specified solar price escalator
      - Discount rate for NPV
      - PPA (or other) contract term in years.

2 Business proposal
   - Legal structure of the developer's company or business organisation
   - Description of experience and capabilities in work of similar scope (e.g. size, complexity, etc.)
   - Advantageous facilities and resources (e.g. technical and operational resources)
   - Familiarity with land lease agreements (agreement will be negotiated once proposal has been selected)
   - Pricing and revenue details
   - References from clients of whom the developer has previously provided services of similar scope
   - Description of proposed subcontractors and their responsibilities.

3 Technical proposal
   - Overview of the proposed method for completing the project and performing the requested services
   - Detailed description of the project approach including the method by which requirements of the scope will be achieved
   - Description of project personnel including experience, qualifications, training, continuing education and certifications of project managers, principal engineers, partners, etc.
   - Identification and mitigation of anticipated potential problems
   - Description of unique features or qualifications the developer can offer
   - Conceptual design and engineering documents providing the following information:
     - System description
     - Equipment details and description
     - Layout of installation
     - Layout of equipment
     - Selection of key equipment
     - Specifications for equipment procurement and installation
     - All engineering associated with structural and mounting details
     - Performance of equipment components and subsystems
     - Integration of solar PV system with other power sources
     - Electrical grid interconnection requirements
     - Controls, monitors and instrumentation
     - Web-based performance monitoring system
     - Foundation of PV support system
   - Provide information regarding location of boundary and system layout to ensure:
     - ease of maintenance and monitoring
     - efficient operation
     - low operating losses
     - secured location and hardware
     - compatibility with existing facilities
   - System monitoring plan
   - Warranties and guarantees
   - Environmental impact statement, environmental management agreement and pollution control plans.

4 Pricing and revenue proposal
   - Estimated costs associated with the development and completion of the project
   - Proposed ownership structure of the project
   - Proposed revenue to the owner, utility, and to the local area.

Proposal evaluation

The application must be submitted by the specified date and will be evaluated based on a set of criteria which may include but is not limited to:

- Overall quality of the proposal
  - structure and organisation
  - clarity
  - completeness and comprehensiveness
  - cohesion and coherence
- Adherence to the specific RFP requirements
  - e.g. size, location, technology type
- Quality of the design
  - operational efficiency
  - environmental impacts
  - utilisation of existing roads and infrastructure
  - compliance with relevant standards, guidelines and regulations
  - utilisation of available area (e.g. co-use of land, tracking technology)
- Financial Strength
  - low, but realistic project costs
  - access to capital
  - good options for sources of finance including debt and/or equity finance and grants
- Experience
  - relevant to project scale and type including design and planning of large-scale solar systems
  - good reputation based on previous successful projects.

Development approval

While some stages of the development approval will be carried out in the tendering process, once a developer has won the contract for a project, they must receive approval to develop the project from a range of stakeholders and government bodies before the project can proceed. Failure to get approval may result in the project being offered to the next best bidder. This can be a very lengthy process depending on a range of factors including the regulatory and political environment, stakeholders involved, complexity of the project, and characteristics of the site, etc.

Requirements for development approval may vary between countries, states and districts but the following permits, licences and agreements are typically required:

- land lease contract
- environmental impact assessment
- building permit/planning consent
- grid connection contract
- power purchase agreement
- easement contract
- development report.

In order to receive these permits and licences, it is necessary to consult with various stakeholders and authorities. This may also vary between countries and regions but may include:

- local and/or regional planning and development authorities
- environmental agencies/departments
- archaeological agencies/departments
- civil aviation authorities (if located near an airport)
- local communities
- landholders
- health and safety agencies/departments
- electricity utilities.

Land use agreements

One of the first stages in planning is to ensure that the rights to construct, use and maintain the land are secured. Lease agreements can be ideal for utility-scale solar projects as they can give the developer the right to use the land without disturbance from the landowner who may not use the land at all or use it only for grazing or minor agriculture.

Solar farms can be land intensive and may leave little open space for dual use of land. This often motivates the land user to limit the amount of land available for use on the lease agreement. This can be problematic for a developer who wants to maximise space for flexibility in layout design, have sufficient space to allow for module cleaning, and clear ground between the perimeter fence and the PV arrays in order to protect them against vandalism from thrown objects.

An alternative to lease agreements is an easement. This means that the land is shared with the landowner. The developer does not have exclusive right to the land, but is given specific use-rights to the property by the landowner. Key components of an easement include:

- A specific term for the lifetime of the solar farm (20–30 years).
- A right to install fixtures and system equipment that remain the property of the developer.
- A clearly defined scope allowing the developer's right to access the land regardless of other facilities which may cross the boundary of the easement.

Quite often, solar farms in the US are developed on land owned by the federal or state government or other bodies such as US Native Americans. Each country's state and federal governments will have different schemes for leasing or licensing public land. The American Bureau of Land Management (BLM) has particular regulations referring to procedures for obtaining land rights, known of ROW grants, discussed previously in the section on regulations and permitting. In the case of Native American tribal land, additional approval from the Bureau of Indian Affairs is necessary before construction can commence.

Development contracts

Development contracts are required between the developer/owner and contractor to ensure that all stages of project development are undertaken successfully and on time. Figure 3.27 illustrates the various agreements that fall under development contracts.

These contracts are often signed under one agreement known as an engineering, procurement and construction (EPC) contract. This can be more

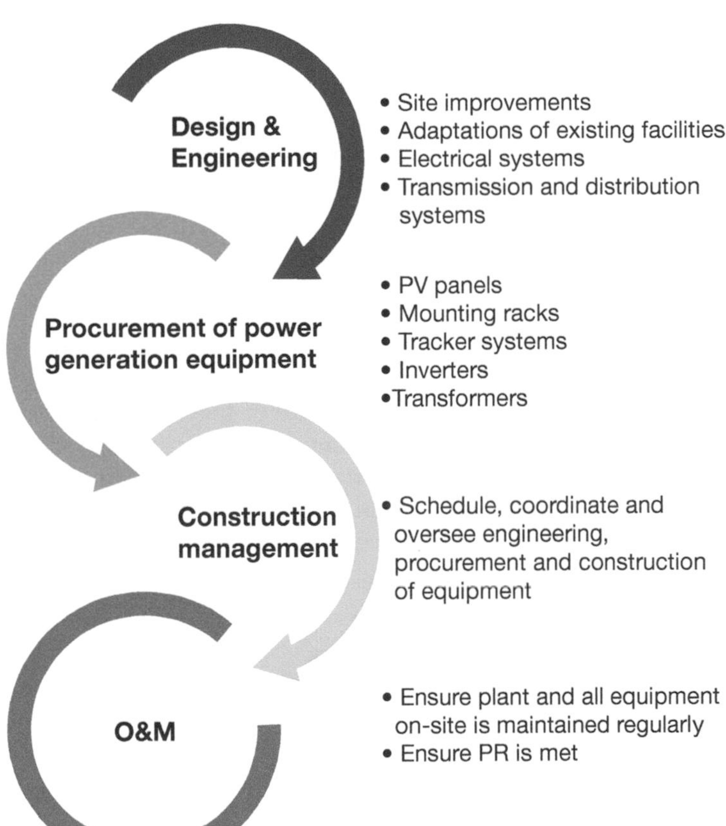

Figure 3.27 Various agreements in a development contract.

expensive but allows for greater integration of all aspects of development as they are being carried out by one entity. The contract will typically be awarded based on experience, proposed price and quality of the proposal.

The other option is for the developer to enter into multiple agreements with various contractors. This can be cheaper and allows the developer to have a direct relationship with key project players. However, it is very important to ensure that each agreement is coordinated effectively to ensure the project is completed on time and on budget.

Power purchase agreement

A power purchase agreement (PPA) is one type of contract used to implement utility-scale solar projects. It involves a seller, typically the developer of the project, and an off-taker (electricity buyer), who is usually a utility, but the off-taker can also be a retailer or end-user such as a manufacturing facility interested in using renewable energy. The off-taker agrees to purchase electricity generated from the solar farm at an agreed price for an agreed duration. The price should be high enough to repay the capital costs of the project, and the contract period should extend beyond the debt amortization period in case there are troubles with finance in later years. This price is typically based on the expected LCOE of the PV plant, calculated by dividing the total expected system and operating costs by the total expected generation in kWh over the contract's period or lifetime of the plant.

PPAs reduce the project risk by guaranteeing a source of income to cover short and long-run marginal costs and hence allow for easier finance. There are many terms and conditions that must be considered and agreed upon for a PPA contract, some examples of contract terms and conditions that may exist are given in Figure 3.28.

Public–private partnership

A public–private partnership (PPP) is a framework that engages both government entities such as departments, ministries and state-owned enterprises with private partners, such as businesses or investors with technical or financial expertise in utility-scale solar. The aim of PPPs is to utilise the relative merits of both the private and public sector in performing certain tasks. For example, the government may be good for capital investment and securing political support, whereas the private partner may contribute expertise in development or EPC, and may also contribute to capital investment.

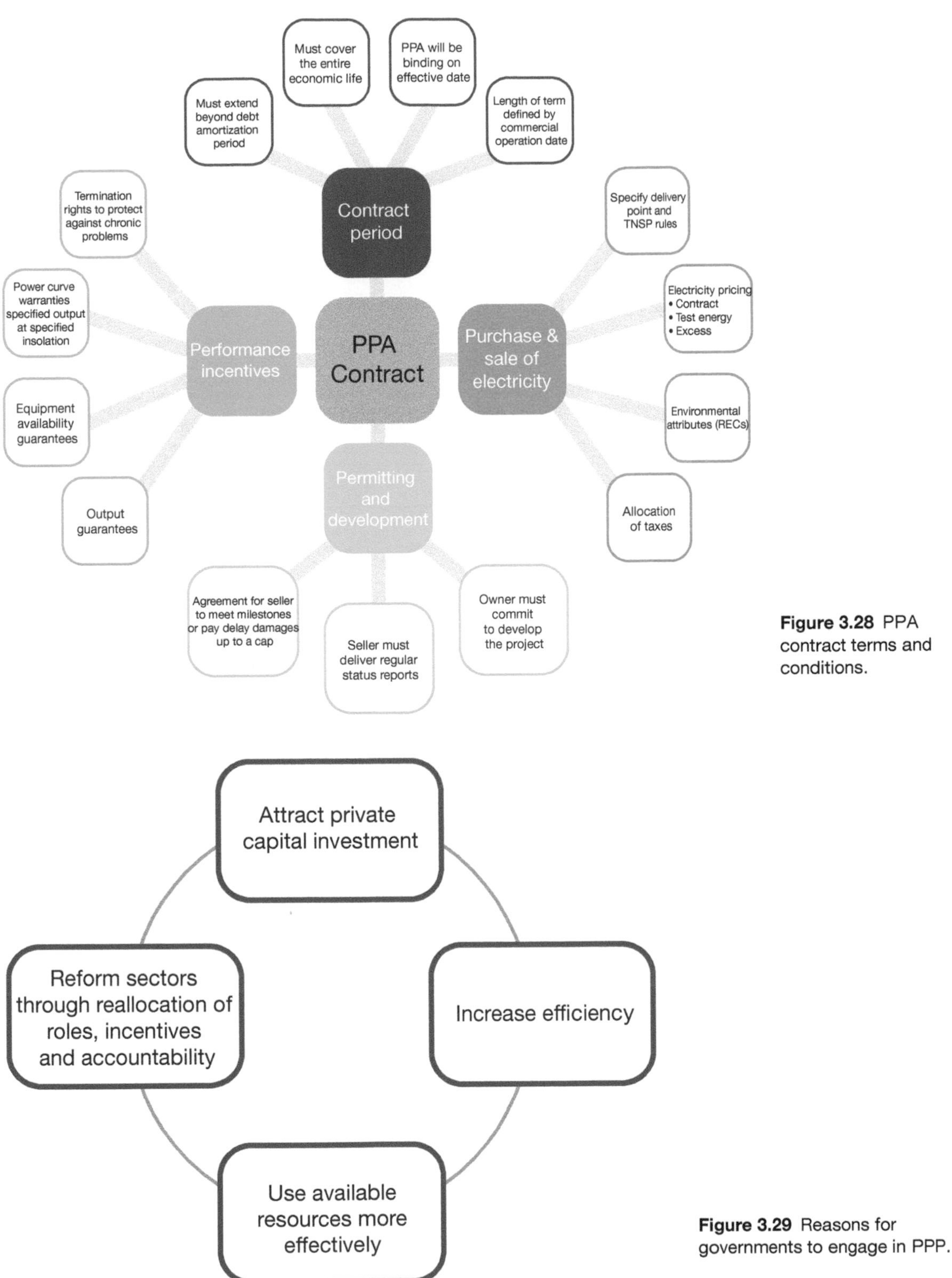

Figure 3.28 PPA contract terms and conditions.

Figure 3.29 Reasons for governments to engage in PPP.

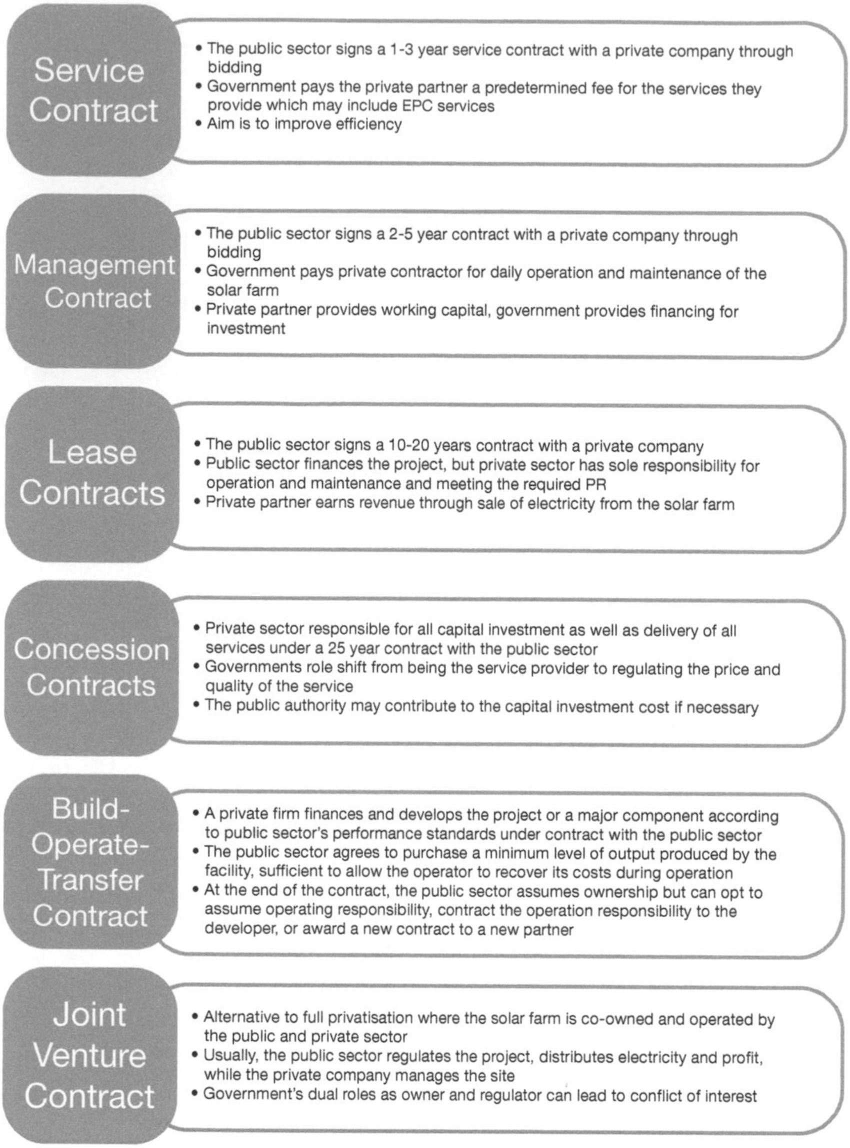

Figure 3.30 Contract options for PPPs.

Source: Asian Development Bank

Community engagement

Stakeholders

Engagement with stakeholders is critical during the early stages of project development to ensure smooth deployment of the solar farm in the local area. It is important to make sure that communication with stakeholders is maintained so that they are always informed and kept up to date with the project and how they may be affected. The key stakeholders will vary from project to project, but it is important to identify them as soon as possible and develop a stakeholder database so that future communication can be easily maintained. Key stakeholders may include but are not limited to:

- local council
- government agencies and regulators
- immediate neighbours
- local and regional community groups
- employment networks
- local businesses
- traditional owners
- landowners
- education providers
- indigenous groups
- media.

Communication strategies

Effective communication strategies will allow for a relationship with stakeholders that is based on mutual trust and respect. This can be achieved through project transparency and regular communication throughout the lifetime of the project. See Figure 3.31 for a list of example strategies that could be used throughout different stages of the project life cycle.

Pre-Construction

- Direct notification
- Meetings and presentations
- Set up community reference group
- Develop marketing materials
- Set up and maintain complaints/comments register
- Participate in local events
- Liaise with employment networks
- Strategic media placement
- Meet with neighbours to discuss impact mitigation
- Set up information website

Construction

- Host start of construction celebration
- Adhere to planning permit
- Update marketing materials regular newletter/construction update
- Participate in local events
- Celebrate key milestones
- Strategic media placement
- Maintain meetings and presentations
- Maintain complaints/ comments register

Operation

- Celebrate completion of construction with community
- Ensure compliance with all relevant legislation and adhere to planning permit
- Ongoing participation with community
- Direct notification to stakeholders
- Be available for one-on-one meetings with stakeholders
- Strategic media placement
- Maintain meetings and presentations
- Maintain complaints/ comments register
- Host site tours

Figure 3.31 Strategies for stakeholder engagement throughout the project lifetime.

Source: FRV

Bibliography

Asian Development Bank. *Public–Private Partnership Handbook*. 2008.

Belfiore, F. *Risks and Opportunities in the Operation of Large Solar Plants*. San Diego: Solar Power-Gen, 2013.

Bell, A. J., M. W. Christopherson, K. G. Erickson, R. M. Faccinto and H. E. Susman. 'Solar Project Property Rights: Securing Your Place in the Sun'. In *Lex Helius: The Law of Solar Energy*, by Stoel Rives, 1–8. 2013.

BRE National Solar Centre. *Biodiversity Guidance for Solar Developments*. Eds G. E. Parker and L. Greene, 2014.

BRE National Solar Centre. 'Planning Guidance for the Development of Large Scale Ground Mounted Solar PV Systems'. 2013.

Department for Communities & Local Government. *Generic Risk Assessment 5.1 – Incidents Involving Electricity*. London: TSO, 2013.

Department of Sustainability, Environment, Water, Population and Communities. 'EPBC Act – Environment Assessment Process'. 2010.

Dillon Consulting. *Sandringham Solar Farm Draft Construction Plan Report*. Invenergy, 2011.

Einowski, E. D., M. A. Lund and H. E. Susman. 'Project Finance for Solar Power Projects'. In *Lex Helius: The Law of Solar Energy*, by Stoel Rives. 2013.

Frankfurt School–UNEP Centre. 'Global Trends in Renewable Energy Investment 2015'. 2015.

Friedman, B., K. Ardani, D. Feldman, R. Citron and R. Margolis. *Benchmarking Non-Hardware Balance-of-System (Soft) Costs for U.S. Photovoltaic Systems, Using a Bottom-Up Approach and Installer Survey – Second Edition*. IREC, 2013.

FRV. 'Community Consultation Plan'. 2014.

FRV. 'Lessons Learned in the Development of Moree Solar Farm'. 2014.

FRV. 'Moree Solar Farm Environmental Assessment Report'. 2011.

GSES. 'Chapter 3 – Solar Radiation'. In *Grid-Connected PV Systems: Design and Installation*, 27–51. Sydney, 2015.

GSES. *Grid-Connected PV Systems: Design and Installation*. 2015.

International Finance Corporation. 'Utility Scale Solar Power Plants'. 2012.

IRENA. 'REthinking Energy'. 2014.

Lawson Fairbank. *Agricultural Land*. n.d. www.lawsonfairbank.co.uk/agricultural-land.asp (accessed 3 August 2015).

Lund, M. A. and J. H. Martin. 'Power Purchase Agreements: Utility-Scale Projects'. In *Lex Helius: The Law of Solar Energy*, by Stoel Rives. 2013.

Maloewitz, Jim. 'Sheep Power: Texas Solar Farm Employs Lamb Landscapers'. *The Texas Tribune*, 7 November 2014.

Merkle, Alan R., Brian J. Nese, and David T. Quinby. 'Solar Energy System Design, Engineering, Construction, and Installation Agreements'. In *Lex Helius: The Law of Solar Energy*, by Stoel Rives. 2013.

Movellan, J. 'Getting Out of the Weeds: How to Control Vegetative Growth under Solar Arrays'. *Renewable Energy World*, 2014.

Nelmes, N. 'Greenough River Solar Farm Community Support Initiative'. n.d. www.greenoughsolarfarm.com.au/node/39 (accessed 28 July 2015).

NGH Environmental. *Preliminary Environmental Assessment: Nyngan Solar Farm*. Infigen Energy, 2010.

NREL. 'A Deeper Look into Yieldco Structuring'. 2014. https://financere.nrel.gov/finance/content/deeper-look-yieldco-structuring (accessed 21 August 2015).

NREL. 'Continuing Developments in PV Risk Management: Strategies, Solutions, and Implications'. 2013.

NSW Government. 'Map Projections'. n.d. www.lpi.nsw.gov.au/surveying/geodesy/projections (accessed 11 August 2015).

NSW Rural Fire Service. 'Planning for Bushfire Protection'. 2006.

OpenEI. 'Transmission Siting & Interconnection'. n.d. http://en.openei.org/wiki/RAPID/Solar/Transmission_Siting_%26_Interconnection (accessed 20 August 2015).

Ryor, J. N. and L. Tawney. *Utility-Scale Renewable Energy: Understanding Cost Parity*. World Resources Institute, 2014.

Sandia National Laboratories. 'Solar Resource Assessment'. http://energy.sandia.gov/energy/renewable-energy/solar-energy/photovoltaics/solar-resource-assessment (accessed 27 July 2015).

SEIA. 'Siting & Permitting'. 2012. www.seia.org/policy/power-plant-development/siting-permitting (accessed 8 August 2015).

SEIA. 'Utility-Scale Solar on Federal Lands'. 2013.

SMEC. 'Valdora Solar Farm Geotechnical Investigation'. 2014.

SolarGIS. 'GeoModel Solar'. 2015. http://solargis.info/doc/free-solar-radiation-maps-GHI (accessed 4 August 2015).

Solar PEIS. 'Solar Energy Development Environmental Considerations'. n.d. http://solareis.anl.gov/guide/environment/index.cfm (accessed 4 August 2015).

US Department of Energy. 'Soft Costs of Solar Deployment'. 2014.

4

Design overview

There are two distinct, but interrelated, components relating to the information and processes necessary to establish a successful solar farm system design: system design and system planning. Both processes are of equal importance and impact the success of the system's financial and performance outcomes.

The system design component will typically comprise the assessment of the proposed site and the operating characteristics of the proposed solar system, based on:

- the equipment proposed for the solar farm (up to and including the point of interconnection with the grid);
- the system layout/electrical design;
- the calculated system performance based on the agreed system performance probabilities.

The system planning assessment has been outlined in Chapter 3 and includes the information that underpins the relevant physical, legal, social, environmental and financial criteria related to the solar farm project. The planning assessment will evaluate the compliance of the proposed solar farm in relation to the system's potential generation. This will be based on the proposed equipment, the system's ability to meet the specified financial and power thresholds and the documented compliance with prevailing standards and guidelines.

The information in this chapter sets out the criteria for the system design component. Where that information is directly related to information given in Chapter 3, it will be indicated. The processes described in this chapter are intended as a guide only. It is the responsibility of the parties undertaking the solar farm design process to ensure that the project meets all proposed performance and financial criteria.

Site assessment

System sizing

One of the very first stages of the design process is to determine the most appropriate system size. This will depend on the amount of capital available to build the project, the amount of available space to accommodate the solar farm, the local electricity demand, the ability of the local grid to accept the output capacity of the solar farm to be injected into the grid network, and incentives that favour projects of particular sizes. Larger systems installed over large areas will have high transmission losses, but these can be mitigated by the use of step-up transformers to minimise these losses through electrical conductors. There will be economies of scale able to be applied to minimise the system costs. The system should be designed to minimise the levelised cost of energy (LCOE) and maximise income from the solar farm.

System arrangement/layout

Once the size of the system is determined, it is important to decide how the proposed components should be best configured on the site. This layout plan will include the dimensions of the array and row-spacing as well as the location of inverters, transformers, cables, combiner boxes and the substation (grid-connection point).

It is usual that a number of different system configurations and layouts will be proposed for consideration during the early stages of the design process. As this process continues, the system design will evolve according to information relating to the site and the system components. See Figure 4.1 for some examples of solar farm layout designs.

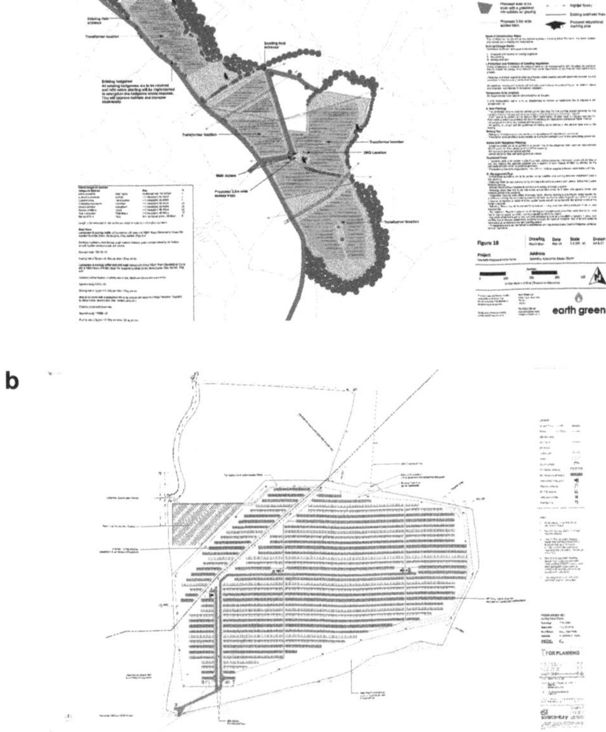

Figure 4.1 (a) Layout design for the 6.5 MW Sawmills Solar Farm (UK). It details the environmental features at the site, showing the location of modules and transformers; landscape and ecology area and buffer; area for grassland and grazing; area for the access track; the location of the deer fence; existing hedgerow; CCTV cameras; existing overhead lines. (b) This layout of the 6 MW Ashby Solar Farm (UK) focuses on the technical details of the solar farm. It includes the location of modules and transformer stations, as well as the location of temporary construction access tracks; shaded areas; permanent O&M access tracks; location of 11 kV line and connection point; site boundary; DC trench; HV trench; water main; CCTV camera; wind sensor pole; satellite pole.

Source: Solstice Renewables

The plan should be drawn up using software such as AutoCAD or other software that has been developed specific to the PV industry such as HELIOS 3D (discussed later in the section on computer-aided design optimising). The system layout will depend on the information gathered in the site assessment during the planning stage (detailed in Chapter 3). This includes:

- shape and size of the available area;
- site boundaries;
- appurtenant structures;
- site access plans; and
- current land use.

Important aspects to consider when designing the layout include:

- Row spacing: should be designed to minimise inter-row shading and the losses associated as well as allow access for maintenance.
- Cable runs: lengths of cable runs should be chosen to reduce electrical losses
- Tilt angle: should be chosen to optimise annual energy yield according to the site latitude and solar resource.
- Module orientation: modules should face the direction that maximises yield (south in the northern hemisphere, north in the southern hemisphere).
- Visual impact: should be minimised e.g. following the contours in the topography of the land.

HELIOS 3D has functions that use this information to plan the layout so that the array can be split into different sections separated by a given distance so that access roads can be built. Areas for inverter and transformers can also be set. The software can also be used to show when and to what extent internal or external objects may cause shading on the array (Figure 4.2). These areas can then be marked as 'keep out' zones, so that no modules are placed in these areas (Figure 4.3).

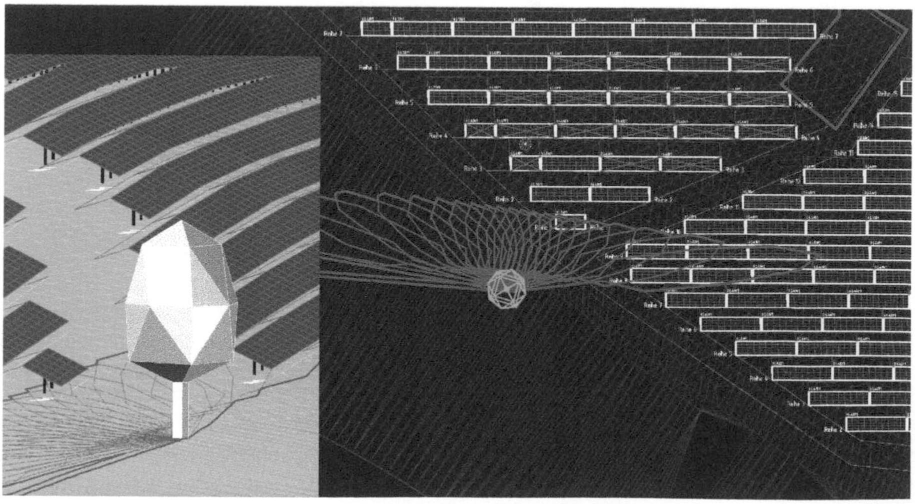

Figure 4.2 HELIOS 3D has many predefined scalable (dynamic) templates that can be used to replicate most shadow casting objects. This image shadow cast by a tree at different times throughout the day.

Source: Stoer+Sauer

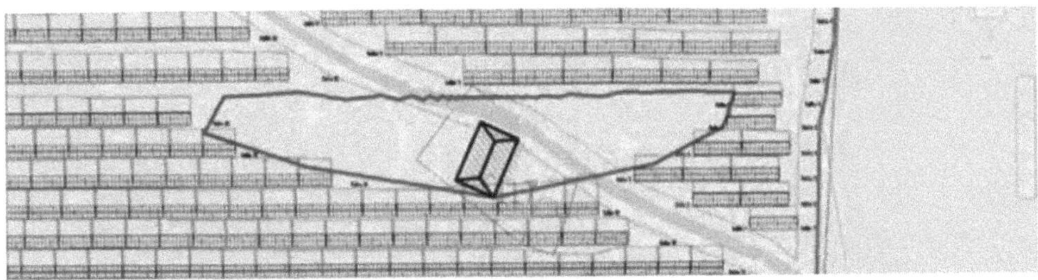

Figure 4.3 HELIOS 3D software can design the layout around the shadow area cast by this object. The area inside the red lines is a 'keep out' zone.

Source: Stoer+Sauer

After structuring the area, it is easy to generate and compare different layouts in order to find the most economical solution. HELIOS 3D can be used to generate different layouts and then import the information to a program such as PVsyst, which can be used to make bankable yield calculations.

Design calculations

Solar resource and radiation

As outlined in Chapter 2, there are site-specific parameters and characteristics that have an impact on the available solar irradiation, which varies widely depending on location (and time of year). When designing a solar farm, it is very important to consider the solar radiation characteristics for the specific location where the solar array will be installed. For instance, a solar farm built to supply electricity to a German city would have to be significantly larger than a solar farm built to supply the same amount of energy to a city in California, USA (see Figure 4.4 for a comparison of the solar resource between Germany and the USA).

Furthermore, locations far from the equator such as Poland receive a large amount of irradiation during long summer days but very little during wintertime when the days are very short. Therefore, in order to maximise annual yield from a solar farm built in this location, the designer may choose to orient the modules so that they face the sun directly during the summer months, instead of during the equinoxes in order to capture abundant solar resource during the long summer days.

In Chapter 3, reference sites are quoted from which site-specific solar irradiation data can be obtained. It is imperative that any quantified solar radiation is based on having applied an approved source of solar irradiation data in relation to the specific installation site for the solar farm. The most obvious requirement for a solar farm is a site for which there is solar PV resource information available.

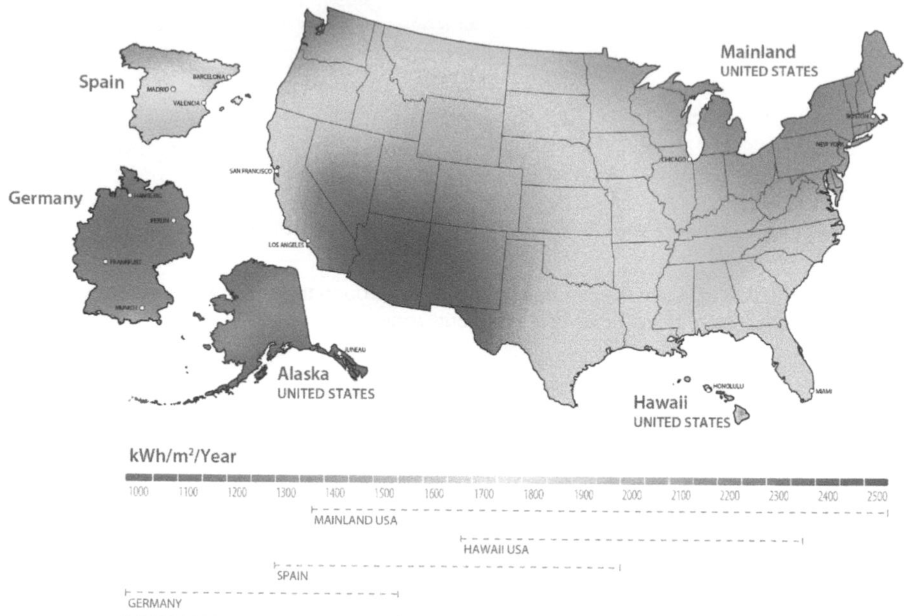

Figure 4.4 This image shows average solar insolation across the USA, Germany and Spain. Power output will be significantly higher in areas with higher solar insolation.

Source: Global Sustainable Energy Solutions

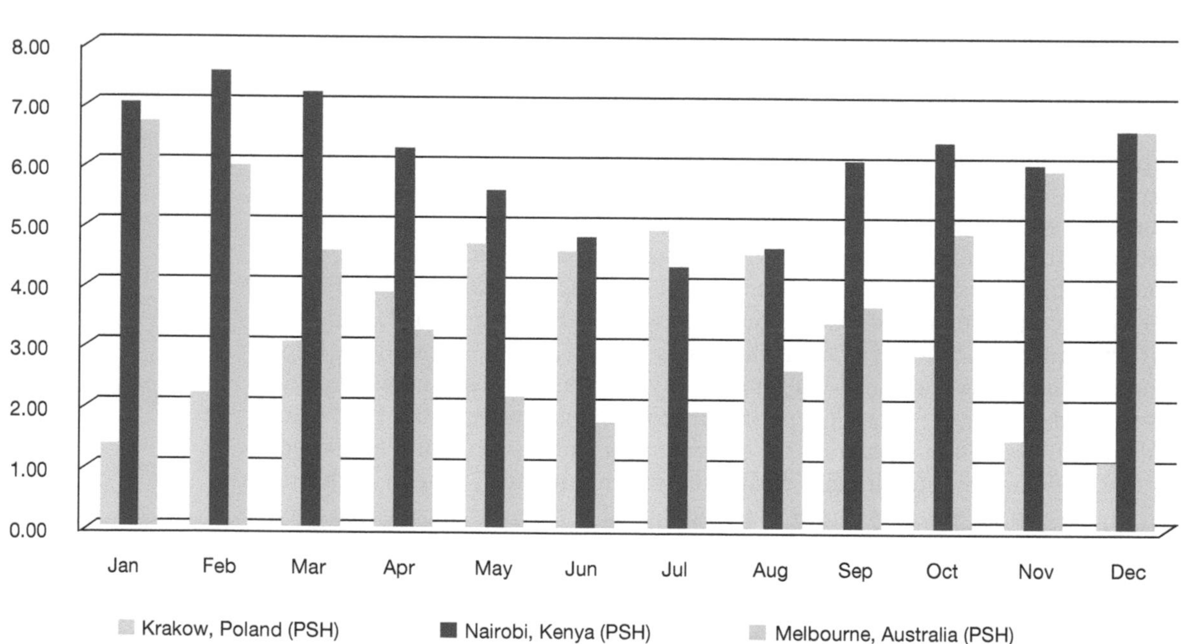

Figure 4.5 Average annual solar insolation for Krakow, Poland (northern hemisphere), Nairobi, Kenya (near the equator) and Melbourne, Australia (southern hemisphere) on a horizontal plane in PSH. This figure shows that areas around the equator (Kenya) receive significantly more solar radiation than areas closer to the poles (Poland and Australia). Areas far from the equator also experience a larger variation in the amount of radiation they receive.

Data source: PVGIS

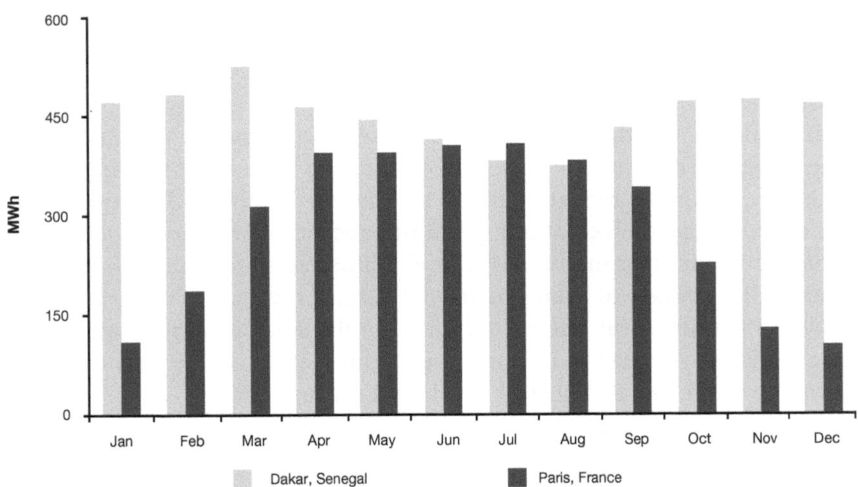

Figure 4.6 Graphed data from PVGIS showing the power output (in PSH) for two identical 100 MWp systems, one installed in Dakar, Senegal and the other in Paris, France. Each installation is installed at the optimum tilt angle as determined by PVGIS (35° from the horizontal for Paris and 17° from the horizontal for Dakar). The installation in Dakar produces more electricity because the solar resource is better. Note that this is a very simplified example, and does not consider site-specific losses, or system design losses. Instead it assumes 14% system losses for both locations.

Data source: PVGIS

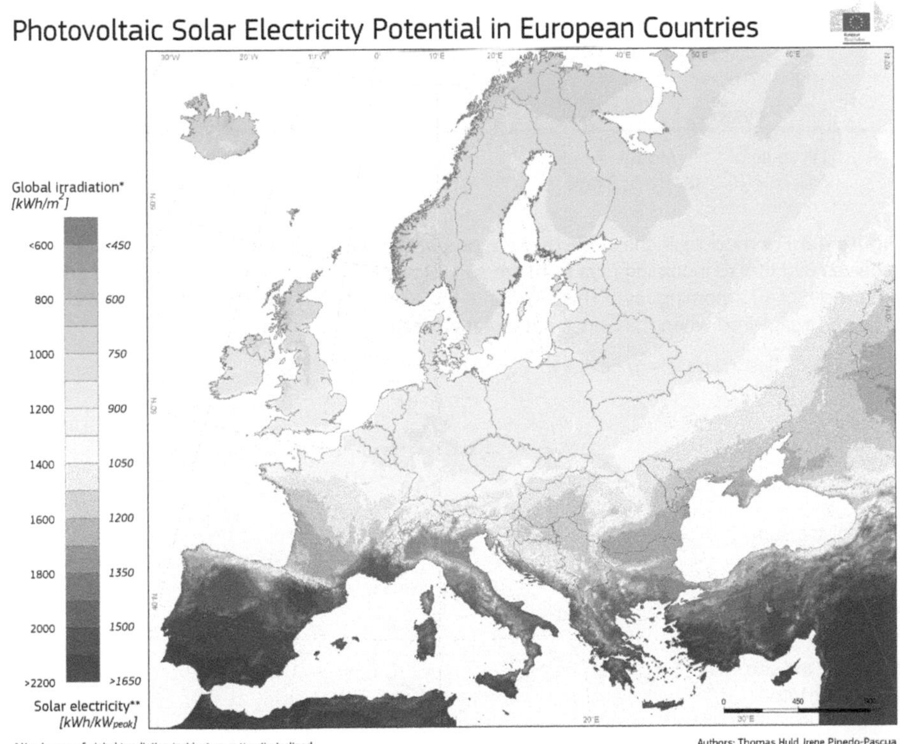

Figure 4.7 This image shows the variation in global radiation across Europe, the southern parts of Europe receive significantly more sunlight. Solar farms located in these areas are likely to have a higher yield than those located in the northern parts of Europe.

Source: PVGIS © European Union, 2001–2012

Tilt angle and orientation

The optimum tilt angle and orientation of solar modules is unique to each site and must be carefully determined to deliver the optimum energy output of the solar farm. Based on the solar resource assessment conducted during the planning stage, the optimal tilt angle and orientation for a solar module can be calculated to maximise the amount of irradiation received. The maximum amount of radiation received at any given time occurs when the sun's rays are perpendicular to the solar modules. Figure 4.8 illustrates the impact that the tilt angle has on the amount of solar radiation a surface receives.

In order to maximise the energy output from the solar farm throughout the year, the modules will need to be positioned at an angle to allow as much solar radiation as possible on average to be collected throughout the year.

As a rule of thumb, in order to ensure that the solar modules will receive the maximum solar radiation over the year, they should be tilted so that they face the sun directly during solar noon on the equinoxes (see Figure 4.9). This is when the altitude of the sun is halfway between the northern and southern solstices (at the equator). This is of course unless vertical-axis trackers are used to follow the sun at different altitude angles throughout the year.

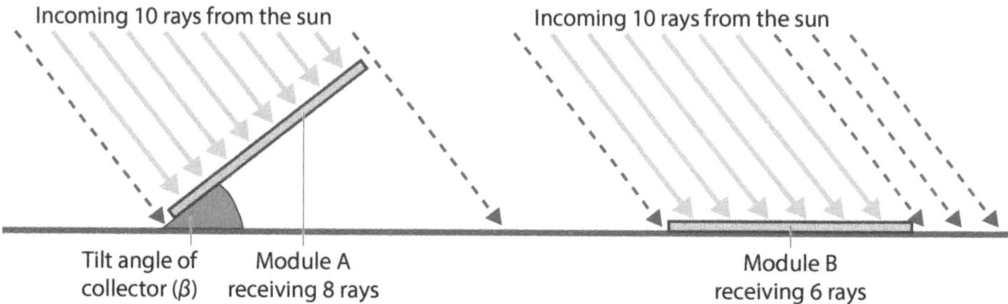

Figure 4.8 For module A, the sun's rays are perpendicular to the surface of the module, and it receives 8 of the 10 incoming rays from the sun. Module B is horizontal to the ground and only received 6 of 10 incoming rays. This highlights the effect of tilt angle on the amount of solar irradiation captured when the Sun is not directly overhead.

Source: Global Sustainable Energy Solutions

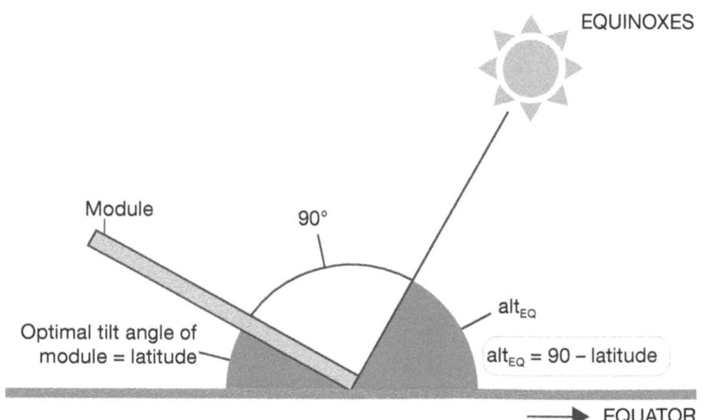

Figure 4.9 Based on the altitude of the sun at solar noon during the equinoxes, the optimal tilt angle of a solar module equals the latitude of the site but takes the opposite cardinal direction. However, this is not necessarily the optimum tilt angle for the site as other factors also need to be taken into consideration.

Source: Global Sustainable Energy Solutions

Box 4.1 The rule-of-thumb equation for calculating optimal module tilt angle

The optimal angle for a PV module at any instant in time is:

Tilt angle = 180° – 90° – solar altitude

As solar altitude at the equinoxes (alt_{EQ}) is required for optimal module annual performance, the equation becomes:

Optimal tilt angle = 180° – 90° – alt_{EQ}

Also, since:

alt_{EQ} = 90° – latitude

Therefore:

Optimal tilt angle = latitude

For example:
Germany is at latitude 52.5°N. Therefore the altitude of the sun during the equinoxes is as follows:

alt_{EQ} = 90° – 52.5° = 37.5°S

Therefore, the optimum tilt angle for a solar module in Germany is given by:

Optimal tilt angle = 180° – 90° – 37.5°
= 52.5° facing south

Note that this is equal to the latitude angle.

In order for the modules to receive direct sunlight throughout the day, they should also be oriented to face the equator (i.e. due north in the southern hemisphere, due south in the northern hemisphere). However horizontal-axis trackers can be used to increase daily solar radiation by following the sun as it moves from east to west throughout the day.

There are exceptions to these rules, for example:

- Modules may not always be oriented optimally, such as when needing to maximise the array's energy production for either the morning or the afternoon or for local, seasonal weather variations.

- Modules should be tilted at a minimum angle of 10° to allow for self-cleaning. This can be an issue in tropical locations where the optimum tilt angle (theoretically) is below 10°.

Computer software is almost always used to calculate the optimum tilt angle based on the solar resource data and specific design requirements.

Inter-row spacing

When calculating the optimum spacing between the rows of solar modules, there may be a compromise required between minimising the inter-row shading and utilising the available land effectively and economically, and minimising the cable runs and resistive losses. There is usually some site-specific shading, for example caused by the module row's shadow length during the early morning and late afternoon, because the sun is low in the sky.

Calculating suitable inter-row spacing utilises the parameters given in Figure 4.10.

$$b = h \times \frac{\cos\lambda}{\tan\alpha}$$
$$h = \sin\beta \times l$$
$$b = \sin\beta \times l \times \frac{\cos\lambda}{\tan\alpha}$$
$$a = \cos\beta \times l$$

Module footprint length at a certain time is then given by:

$$d = a + b = \cos\beta \times l + \sin\beta \times l \times \frac{\cos\lambda}{\tan\alpha}$$

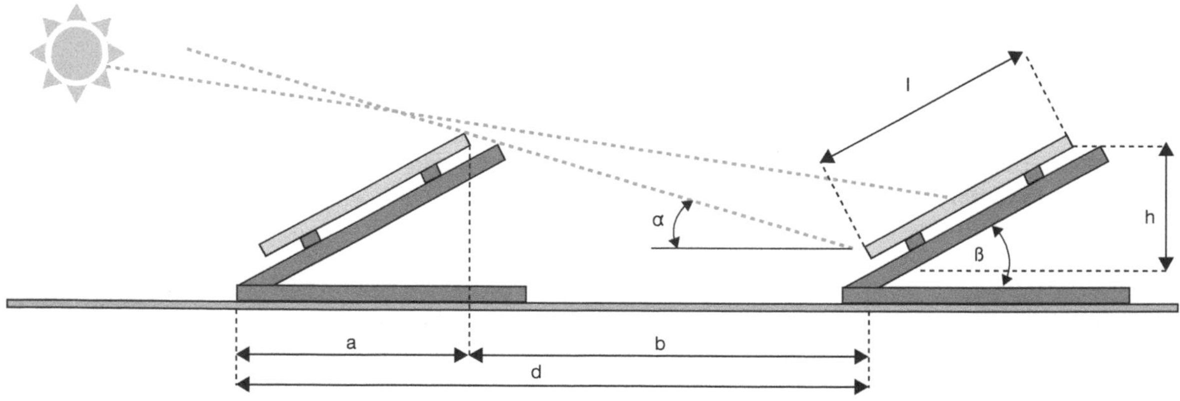

Figure 4.10 Shading angle diagram.

- α = altitude angle: the vertical position of the sun
- λ = azimuth angle: position of the sun with respect to north
- β = module tilt angle: the tilt angle can be reduced as a means of reducing the module footprint length d without increasing the distance between rows
- h = vertical height of modules: used to calculate the length of shadow b
- l = the module length: used to calculate the vertical height of modules h
- a = shadow length under modules: used to calculate module footprint length d
- b = shadow length behind module: the length of the shadow that may be cast on to other modules
- d = module footprint length: total length of shadow cast by module, used to determine the inter-row spacing.

The module footprint length should be calculated throughout the year to minimise shading. As a rule of thumb, the rows of modules should be spaced so that there is no inter-row shading at solar noon on the winter solstice (21 June in the southern hemisphere, 21 December in the northern hemisphere). As a general rule, the rows should be spaced such that there is less than 1% annual loss due to shading.

Software is available to design the layout of the system such that inter-row shading is minimised. HELIOS 3D is a layout tool that can design the system such that there is no shading for an agreed number of hours either side of any modules at any time of the day or year. It does this by using site-specific geographic/cartographic data gathered during the planning stage, combined with a reference date and time to define the actual angle of the sun (see Figure 4.11).

Figure 4.11 To achieve shadow free layout, HELIOS 3D calculates the longest shadow cast in a row, and then sets the distance between those rows to that value.

Source: Stoer+Sauer

Electrical design

Matching the array to the inverter(s)

The PV array will be sized based on the system capacity determined in the early stages of design, discussed in the system sizing section. The configuration of the PV array(s) must be designed to match the inverter's specifications. Selecting the most appropriate size and number of inverters is often an iterative process, and will vary from site to site.

Issues to consider include:

- The economics: having fewer larger inverters is usually cheaper than using multiples of smaller inverters.
- Site conditions: a site that doesn't allow uniform module orientation and tilt angle across the system may need multiple smaller inverters, instead of few larger ones, and may have a higher installation labour cost.
- The accessibility of the site: if system maintenance and servicing is not able to be easily provided, e.g. the site is in a remote location, having several inverters means that if there is an inverter malfunction only one part of the system will be offline. Spare inverters can be kept in stock to enable speedy repairs, which is known as redundancy. The overall system design and economics will address this issue and indicate whether offsetting remote servicing issues by having additional equipment on-site is preferable to having extended 'down time' of all or part of the solar farm while having equipment serviced.
- Matching the array capacity: using only a few (or a single) larger inverters means there is less flexibility when matching inverter capacity with PV array capacity.

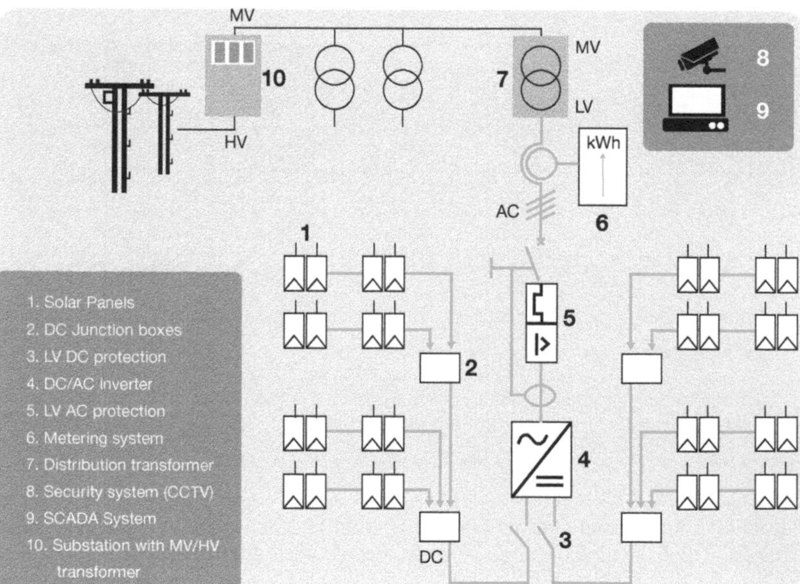

Figure 4.12 Electrical components in utility-scale solar system.

The system design requires that a range of possible configurations and inverter options are examined and a decision is reached regarding the overall system design based on practical and financially sound criteria. The array design can then be refined to suit the specific inverter specifications. Designing the array to match the inverter will require careful consideration of the inverter specifications. If a string inverter is used, the number of modules per string will be determined by the voltage and power limits of the inverter. Central inverters used for utility-scale systems typically have multiple inputs. Each input will be connected to a sub-array which consists of a cluster of parallel strings. The number of modules in series will again be based on the voltage limits of the inverter, but the total number of parallel strings that are connected to the inverter will depend on the inverter current limits. The total number of parallel strings will be split in to sub-arrays that are connected to each inverter input. The sub-arrays will then be connected in parallel on a busbar inside the inverter, which means that there is typically no need for an array combiner box. However, a string combiner box may be required to facilitate string fusing. The number of inverters required to form the whole array will be determined by the total capacity required of the solar farm.

While it is possible to use micro and string inverters for utility-scale solar projects, central inverters are usually the most economic choice and are hence almost always chosen for utility-scale application. This section will describe how to size the array once the appropriate size and number of central inverters have been selected. Although having a large number of modules per string will increase the voltage and hence reduce resistive losses, the maximum voltage allowed per string is limited according to equipment safety requirements, inverter voltage limits and national regulations. The steps for matching the array to the inverter is summarised in Figure 4.13 but will be described in detail throughout this section.

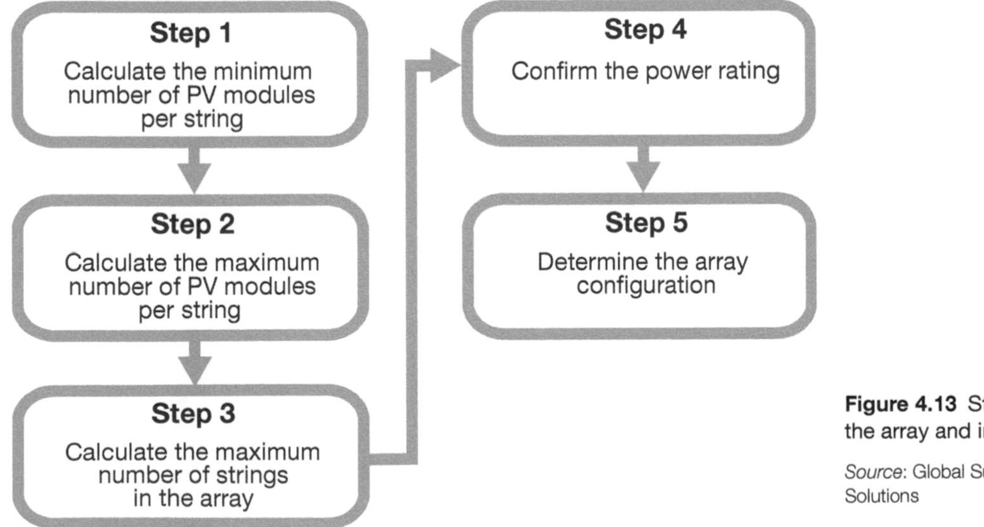

Figure 4.13 Steps for matching the array and inverter.

Source: Global Sustainable Energy Solutions

The calculations used to size the array must include power, voltage and current ratings:

- Voltage ratings: determine the minimum and maximum number of modules in each string.
- Current ratings: determine the maximum number of parallel strings in the sub-array and/or array.
- Power ratings: determine the maximum number of modules in the sub-array and/or array.

These ratings are represented visually in Figure 4.14.

The I–V curve of an array varies based on the irradiance and cell temperature of the modules. Therefore, the array must be designed with its expected operating conditions in mind so that it stays within the voltage, current and power ratings of the inverter as the irradiance and temperature change.

Impact of temperature on PV module voltage

Temperature variation has the biggest impact on the voltage at which a PV module or array operates, so voltage limits should always consider the range of temperatures possible at the site. Steps for calculating temperature dependent voltages are as follows:

- Step 1: calculate the difference between the cell temperature (whether it be the maximum, minimum, or average) and the STC temperature (25°C).
- Step 2: multiply the temperature difference by the voltage coefficient in V/°C.

Note:

- If the cell temperature is higher than 25°C, the Step 2 value will be negative, and hence the voltage decreases.

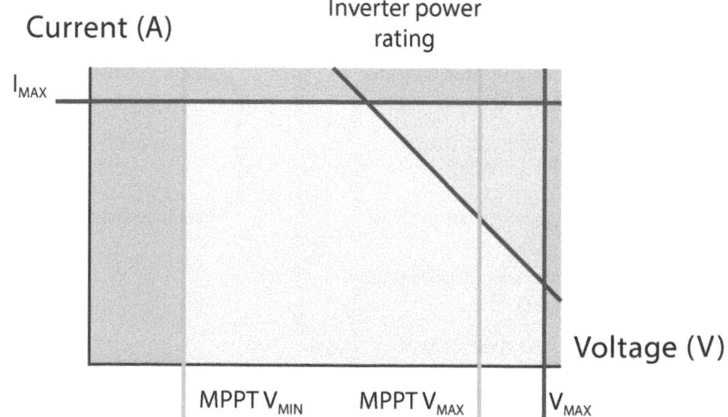

Figure 4.14 Array sizing limits. The blue lines show inverter voltage range that the modules must remain between in order to operate at the maximum power point (MPPT V_{MIN} and MPPT V_{MAX}). The red lines show the maximum voltage and current limit of the inverter (V_{MAX}). The purple line shows the inverter power rating. The green area shows the conditions that the array I–V curve must meet at all times. In some circumstances, it may be possible to configure an array so that it is able to operate safely within the inverter V_{MAX} and I_{MAX} but outside the inverter power rating (the orange area). This is known as oversizing the array and is explained later in this section. The array should not operate in the red area under any circumstances.

Source: Global Sustainable Energy Solutions

- If the cell temperature is lower than 25°C, the Step 2 value will be positive, and hence the voltage increases.

The module voltage for a specified temperature is calculated using the following equation:

$$V_{at\ X°C} = V_{at\ STC} + (\gamma_V \times (T_{X°C} - T_{STC}))$$

Where:

$V_{at\ X°C}$ = voltage at the specified temperature (X°C), in volts
$V_{at\ STC}$ = voltage at STC (i.e. the rated voltage), in volts
γ_V = negative voltage temperature coefficient, in V/°C
$T_{X°C}$ = cell temperature, in °C
T_{STC} = temperature at STC, in °C (i.e. 25°C).

Step 1: Calculating the minimum number of modules in a string

The minimum number of modules in a string is governed by the requirement that the string voltage remains above the inverter's MPPT minimum threshold. If the array or string MPP voltage (V_{MPP}) falls below the minimum operating voltage, the system will underperform, or in the worst case the inverter will shut down. If the array or string open circuit voltage (V_{OC}) falls below the minimum operating voltage, the inverter will turn off.

The calculations applied to determine that the module voltage does not drop below the inverter's MPPT voltage threshold must consider three key factors:

1. Module voltage decreases as temperature increases: the V_{MPP} figure is calculated for the highest expected cell temperature. It is recommended that a minimum of 70°C is used for this temperature, although it may vary depending on specific site conditions; for cool climates, a lower maximum cell temperature could be used.
2. Voltage drop in the DC cabling between the modules and inverter: voltage drop in the DC cabling means that the inverter receives a voltage less than the modules produce. The module's operating voltage output needs to be adjusted to account for this. For example, a voltage drop of 2% means that the module's minimum voltage should be multiplied by 0.98 ((100% – 2%) = 98% = 0.98) to account for a 2% loss.
3. Safety margin for the inverter's minimum voltage: to ensure that the module's voltage does not drop below the inverter minimum voltage, the calculations should provide for a 10% safety margin to the inverter's minimum voltage. This can be achieved by multiplying the module's minimum voltage by 0.9 ((100% – 10%) = 90% = 0.90). This safety margin allows for other factors that may influence the module voltage, such as:
   - the inverter not always operating at the ideal MPP
   - the V_{MPP} voltage decreasing with a decrease in irradiance
   - manufacturing tolerances
   - shading on the array.

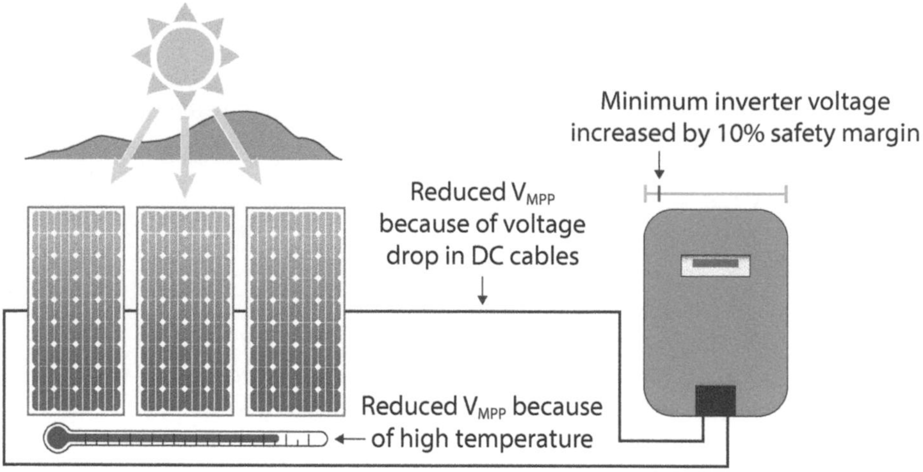

Figure 4.15 Three key factors that should be taken into consideration when calculating the minimum module voltage.

Source: Global Sustainable Energy Solutions

The minimum number of modules (N_{MIN}) can then be calculated using the following equation:

$$V_{MPP(module\ MIN)} = V_{MPP\ at\ 70°C} \times 0.98 \times 0.9$$

$$V_{MPP(module\ MIN)} \times N_{MIN} > V_{MPP\ (inv\ MIN)}$$

$$\Rightarrow N_{MIN} > \frac{V_{MPP\ (inv\ MIN)}}{V_{MPP\ (module\ MIN)}}$$

Step 2: Calculating the maximum number of modules per string

The maximum number of modules in each string is calculated so that the following conditions are satisfied:

- The open circuit voltage (V_{OC}) does not exceed the maximum input voltage: if this occurs, it could damage the inverter.
- The operating voltage (V_{MPP}) should not exceed the MPPT maximum voltage threshold: if this occurs, the array will not operate at its MPP, reducing the power generated. It could be possible to connect more modules in series so long as the resulting V_{OC} will not exceed the maximum input voltage.

For these calculations, the voltage produced by the modules in the early morning is used because this is when the modules will be at their coldest: the air temperature is cool and the modules have not yet warmed up.

To determine the maximum number of modules, calculations are based on three key factors (Figure 4.16):

1 Module voltage increases as temperature decreases: the module's maximum voltage is calculated at the lowest expected cell temperature (during operation), which is most likely at dawn. The minimum cell temperature will vary from site to site.
2 There is no voltage drop in the DC cabling between the modules and inverter at V_{OC}: the calculations are performed at V_{OC}. This means that there will not be any voltage drop, as there is no current flow along the cables.
3 Safety margin of inverter maximum voltages: to ensure that the modules will not exceed the inverter's maximum voltages, it is recommended to include a safety margin of 5% on the inverter maximum voltage. This safety margin is to protect the inverter, to ensure that it will not receive an excessively high voltage.

The maximum number of modules (N_{MAX}) can then be calculated using the following equations:

$$V_{OC\ (module\ MAX)} = V_{OC\ at\ 5°C} \times 1.05$$

$$V_{OC\ (module\ MAX)} \times N_{MAX} > V_{DC\ (inv\ MAX)}$$

$$\Rightarrow N_{MAX} > \frac{V_{DC\ (inv\ MAX)}}{V_{OC\ (module\ MAX)}}$$

$$V_{MPP\ (module\ MAX)} = V_{MPP\ at\ 5°C} \times 1.05$$

$$V_{MPP\ (module\ MAX)} \times N_{MAX} > V_{MPP\ (inv\ MAX)}$$

$$\Rightarrow N_{MAX} > \frac{V_{MPP\ (inv\ MAX)}}{V_{MPP\ (module\ MAX)}}$$

Note: the first equations shown above are important in order to ensure that the inverter is protected. The second equation is less critical because the additional voltage at this time is so small and only lasts for a very short period before the modules heat up.

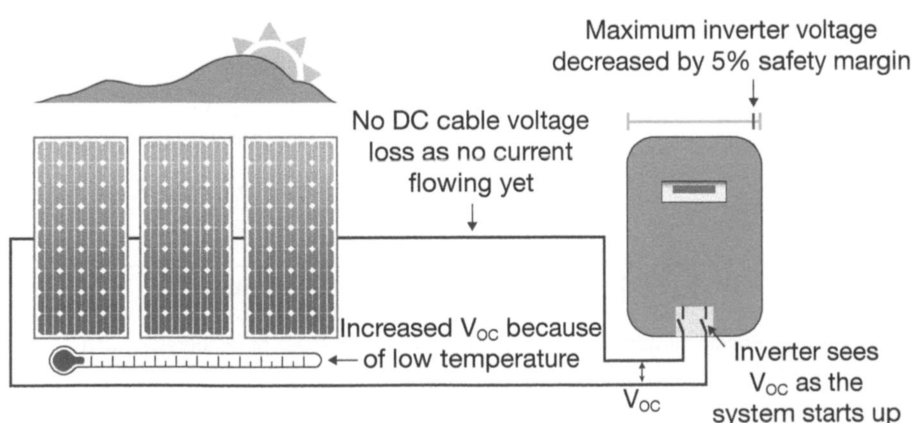

Figure 4.16 Key factors to consider when determining the maximum module voltage.

Source: Global Sustainable Energy Solutions

Step 3: Calculating the number of strings connected to the inverter

The inverter's maximum allowable current will govern the maximum number of parallel strings that can be connected to the inverter. Having the inverter exposed to excessive current can lead to premature aging of the inverter as well as loss of yield.

As with a module's voltage, the effect of external factors on I_{SC} should be taken into consideration:

- Irradiance: a module's current is mostly determined by the irradiance received. Higher irradiance results in higher current (Figure 4.17a). However, the I_{SC} given in a module's datasheet does not need to be corrected because it has already been calculated at an irradiance value of 1 kW/m^2 (STC), which is the maximum irradiance level that will be experienced by most PV arrays. The few situations where the modules receive more than 1 kW/m^2 will be covered by the nominal safety margin included by using I_{SC} instead of I_{MPP}.
- Temperature: a module's current is only slightly affected by the cell temperature: higher temperature results in a slightly higher current (Figure 4.17b). A quick calculation can estimate the maximum number of strings without correcting the I_{SC} for temperature. However, a more thorough calculation using a temperature-corrected I_{SC} should be carried out for utility-scale systems.

To calculate the module maximum current, the following formula can also be used:

$$I_{SC\text{ at }X^{\circ}C} = I_{SC\text{ at STC}} + \left(\gamma_{I_{SC}} \times \left(T_{X^{\circ}C} - T_{STC}\right)\right)$$

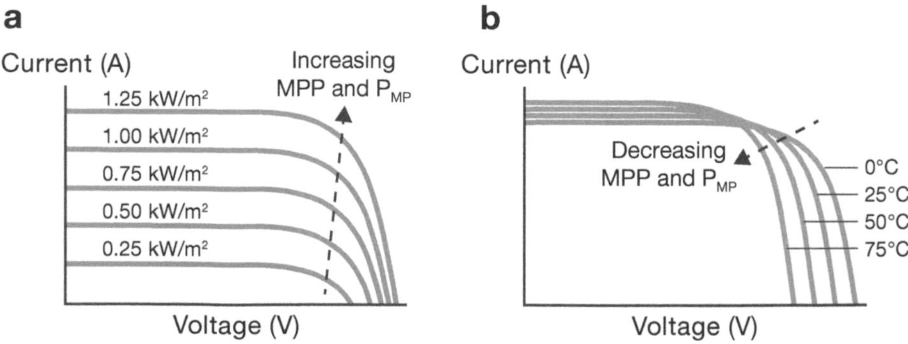

Figure 4.17 (a) Effect of irradiance on a module's current; (b) effect of temperature on a module's current.

Source: Global Sustainable Energy Solutions

Where:

$I_{SC\ at\ X°C}$	= short-circuit current at the specified temperature (X°C) in amperes
$I_{SC\ at\ STC}$	= short-circuit current at STC, i.e. the rated current in amperes
γ_{ISC}	= current temperature coefficient in A/°C
$T_{X°C}$	= cell temperature, in °C
T_{STC}	= temperature at STC, in °C (i.e. 25°C)

Once this current is calculated, the maximum number of strings (N_{MAX}) can be calculated using the following equation:

$$I_{SC\ (module\ MAX)} = I_{SC\ at\ 70°C}$$

$$I_{SC\ (module\ MAX)} \times N_{MAX} > I_{(inv\ MAX)}$$

$$\Rightarrow N_{MAX} > \frac{I_{(inv\ MAX)}}{V_{SC\ (module\ MAX)}}$$

Step 4: Confirming the power rating

Before the optimum configuration of modules and strings can be selected, it is important to ensure that the rated output of the array does not exceed the inverter's rated input power (see Figure 4.18). This is unless the array is oversized, which will be described below.

Note: this may have been one of the first steps when selecting the number and size of inverters for the whole array.

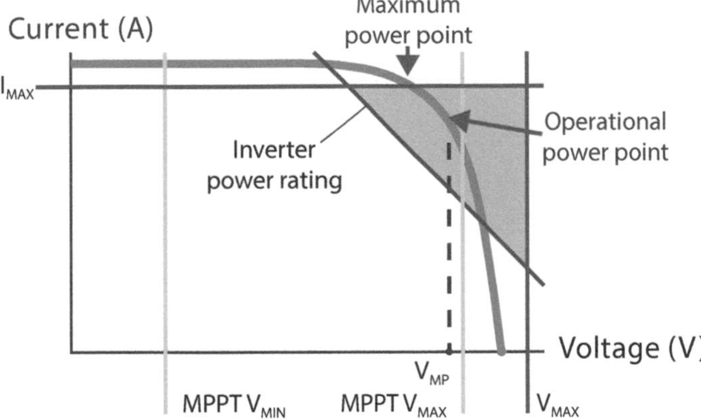

Figure 4.18 The MPP should always sit below the power rating of the inverter. The exception to this is when oversizing the array. Note that the red lines show the maximum current and voltage that the inverter can handle, the light blue lines show the MPP voltage range of the inverter, the purple line shows the rated power of the inverter, and the blue line shows the IV curve of the array.

Source: Global Sustainable Energy Solutions

The maximum number of PV modules in a sub-array connected to a single inverter can be calculated simply by dividing the inverter's rated power by the module's rated power:

$$\text{Maximum number of modules} = \frac{\text{Inverter's rated input power}}{\text{Module's rated output power}}$$

Oversizing the array

Oversizing the array describes the situation where more modules are connected to the inverter than the power-matching calculations allow, but the system remains within the inverter's voltage and current ratings. This is because solar arrays will quite often produce less than their maximum rated power due to many factors, such as the module tilt and orientation, cloud cover, high temperatures, dust, or simply the time of day (e.g. morning or afternoon periods, when there is less irradiation than at solar noon). Oversizing the array will mean that, during those times of reduced PV-array performance, the array's output is able to use more of the inverter's rated power processing capacity. During times of optimum performance by the array, the inverter will manage the extra power by controlling the voltage and current. This will mean that the output is clipped during optimal conditions but can increase the overall output of the system.

Before designing an oversized array, the inverter manufacturer should be consulted to ensure that combining an oversized array with an inverter would not cause the inverter to age faster as it is operating at its maximum power more often. It is also important to ensure that the inverter warranty will not be compromised if the array is oversized. Inverter manuals also often provide guidelines.

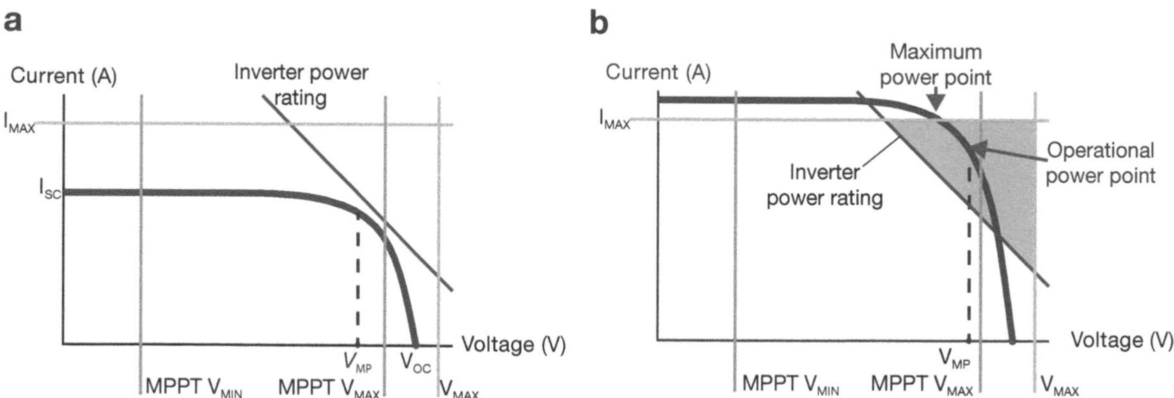

Figure 4.19 (a) Usually the MPP of the array is designed so that it sits below the inverter rating. Figure (b) shows the I–V curve of an oversized array. This option can increase the overall yield of the system by allowing the array output to use more of the inverter's capacity during periods of reduced PV performance. During periods of optimum performance, the inverter will increase the voltage and decrease the current to manage the extra power, ensuring that the inverter's I_{MAX} is never exceeded.

Note: the dark blue line represents the IV curve of the PV array in each figure.
Source: Global Sustainable Energy Solutions

Step 5: Determining the array configuration

After completing steps 1–4, the following parameters will have been calculated:

- minimum number of modules in each string;
- maximum number of modules in each string;
- maximum number of strings connected to the inverter; and
- maximum number of modules connected to the inverter.

Based on this information there is likely to be a number of possible array configurations. Some factors to consider when selecting the most appropriate configuration are:

- number of string combiner boxes/string monitor connections;
- number of string overcurrent protection devices;
- acceptable cable losses;
- number and size of the required DC isolators;
- number of inputs needed for redundancy and maintenance; and
- oversizing factor for optimal return on investment.

Worked example

An example of a design of the 100 MWp grid-connected PV system using central inverters is provided below. This example uses 300 Wp monocrystalline modules (technical data provided in Table 4.1) and the SMA Sunny Central 900CP XT inverter (technical data provided in Table 4.2).

Table 4.1 Technical data for 300 Wp monocrystalline PV module

Electrical characteristic	Value
Maximum power at STC (P_{MAX})	300 Wp
Optimum operating voltage (V_{MP})	32.0 V
Optimum operating current (I_{MP})	9.46 A
Open circuit voltage (V_{OC})	39.5 V
Short circuit current (I_{SC})	10 A
Temperature co-efficient of V_{OC}	–0.31%/°C
Temperature co-efficient of P_{MAX}	–0.42%/°C
Maximum series fuse rating	20 A
Maximum system voltage	1,000 V
Number of cells	60
Power tolerance	0 to +3%

Temperature correction of module voltages

As discussed above the voltage ratings of the modules need to be adjusted for minimum and maximum cell temperatures. For this example, it is assumed that the site has a minimum ambient temperature of 5°C, which means that the minimum cell temperature will be 5°C and the maximum cell temperature is assumed to be 70°C (note that this value is country specific).

Using the principles given above the module voltage ratings have been temperature corrected with a voltage drop percentage (of 3%) added to the V_{MPP} calculations:

$$V_{MPP\text{ at }X°C} = (V_{MPP\text{ at }STC} + (\gamma_V \times (T_{X°C} - T_{STC}))) \times (1 - \text{voltage drop decimal})$$

$$V_{MPP\text{ at }75°C} = (32V + [-0.1344V / °C \times (70°C - 25°C)]) \times (1 - 0.02) = 26.07V$$

$$V_{MPP\text{ at }5°C} = (32V + [-0.1344V / °C \times (5°C - 25°C)]) \times (1 - 0.02) = 34.63V$$

$$V_{OC\text{ at }5°C} = 39.5V + [-0.1225V / °C \times (5°C - 25°C)]) = 40.69V$$

Matching array and inverter calculations

The central inverter design requires each inverter to be optimised and then aggregated to determine system size.

Table 4.2 Technical data for Sunny Central 900CP XT inverter

Input data (DC)	
Max. DC power (at cos φ=1)	1,010 kW
Max input voltage	1,000 V
V_{MPP_MIN} at $I_{MPP} < I_{DCMAX}$	596 V
MPP voltage range (at 25°C/50°C at 50 Hz)	722–850 V/656–850 V
Rated input voltage	722 V
Max input current	1,400 A
Number of independent MPP inputs	1
Number of DC inputs	9
Output (AC)	
Rated power (at 25°C)/nominal AC power (at 50°C)	990 kVA/900 kVA
Nominal AC voltage/nominal AC voltage range	405V/365–465V
AC power frequency/range	50 Hz, 60 Hz/47–63 Hz
Rated power frequency/rated grid voltage	50 H/405 V
Max. output current/max. total harmonic distortion	1,411 A/0.03
Power factor at rated power/displacement power factor adjustable	1/0.9 leading–0.9 lagging
Feed-in phases/connection phases	3/3

Source: SMA Solar Technology

Minimum number of modules in a string (according to $V_{MPPT_{MIN}}$):

$$\text{Min. no. modules in a string} = \frac{\text{min. MPPT voltage} \times \text{safety margin}}{\text{Min. module voltage (at 70°C)}}$$
$$= \frac{596\text{V}}{26.07 \times (1 - 0.1)}$$
$$= 25.4 = 26 \text{ modules (rounded up)}$$

Maximum number of modules in a string (according to inverter limits):

$$\text{Max. no. modules in a string} = \frac{\text{Max. inverter voltage} \times \text{safety margin}}{\text{Max. module voltage (at 5°C)}}$$
$$= \frac{1{,}000\text{V}}{40.69 \times (1 + 0.05)}$$
$$= 23.87 = 23 \text{ modules (rounded up)}$$

From these calculations, it can be seen that there is no suitable number of strings for the selected PV modules and inverter, because the maximum is lower than the minimum number of modules. This is because the voltage range on the inverter is very tight. One option is to try different modules or inverters, but it is the designer's decision whether to include the safety margin assuming the inverter could not be damaged. If the safety margin were not considered, the results would be the following:

$$\text{Min. no. modules in a string} = \frac{\text{Min. MPPT voltage}}{\text{Min. module voltage (at 70°C)}}$$
$$= \frac{596\text{V}}{26.07}$$
$$= 22.86 = 23 \text{ modules (rounded up)}$$

$$\text{Max. no. modules in a string} = \frac{\text{Max. inverter voltage}}{\text{Max. module voltage (at 5°C)}}$$
$$= \frac{1{,}000\text{V}}{40.69}$$
$$= 24.58 = 24 \text{ modules (rounded down)}$$

From these calculations, it is acceptable to have 23 or 24 modules in each string.

Maximum number of parallel strings on each inverter:

$$\begin{aligned}\text{Max. no. strings} &= \frac{\text{Max. inverter current}}{\text{Module short} - \text{circuit current}}\\ &= \frac{1{,}400\text{A}}{10\text{A}}\\ &= 140 \text{ strings}\end{aligned}$$

System configuration

The inverter can have 24 modules per string, with a maximum of 140 strings on each inverter. Each inverter has nine DC inputs, each connected to a sub-array. Since 140 strings cannot be evenly split into nine inputs, each sub-array will not have the exact same number or strings. One option is to have four sub-arrays made up of 15 strings, and five sub-arrays made up of 16 strings, which totals 140 strings connected to the inverter. The rated DC input of the inverter is 1,010 kW, which means that a 100 MWp array will require 100 of these central inverters. Therefore, this system will be designed to have 100 central inverters, each connected to 140 strings made of 24 modules each. This will give total array power:

$$P_{array} = 100 \text{ inverters} \times 140 \text{ strings} \times 24 \text{ modules}$$
$$P_{array} = 100.8 \text{ MWp}$$

Note that 139 strings connected to each inverter would actually give almost exactly 100 MWp, so if this turns out to be more economic, or if space is limited, this option may be more suitable. However, more of the inverter's capacity is utilised if 140 strings are connected, as opposed to 139, but only by a small amount.

Transformer's sizing

For a utility-scale PV plant, there will usually be a transformer installed with each inverter (known as a 'distribution transformer') and a larger transformer at the substation (known as the 'substation transformer'). Sizing these distribution transformers can be challenging. Many manufacturers recommend to size distribution transformers at the same power rating as the inverters they are connected to. This is because high temperatures can shorten the lifetime of the transformer, so it is important not to overload it. Oversizing transformers also means that they will operate at a reduced efficiency for long periods of time. On the other hand, using transformers that are too small prevents utilisation of the plant's maximum capacity.

The most appropriately sized transformer can be selected on the basis of the relative costs associated with different transformer sizes and determining the optimum, financial solution. Costs to be included will include:

- initial cost;
- cost of the energy wasted due to transformer overloads;

- cost of energy wasted due to transformer efficiency; and
- cost of energy wasted due to grid instability.

Transformers should be subject to a number of routine and type tests performed on each model manufactured; these tests are set out in IEC/TS 60076.

Fault prevention

Circuit protection must be applied to all electrical systems to ensure their safe operation to protect equipment and people against faults. It is important to consult relevant local and international standards, local service rules and the network operator's regulations when designing the overall protection system.

The system protection required for a grid-connected PV system includes:

- Overcurrent protection: automatically triggers when a fault current is registered, disconnecting the relevant components. This device protects the system against the high current levels experienced during a short circuit or system overload.
- Disconnection devices: used to manually switch parts of the PV system on or off, as required, during emergency events and maintenance. Disconnection devices in a grid-connected PV system are used for both DC and AC circuits, and some specific disconnection devices are mandatory for a grid-connected PV system.
- Earthing/grounding: prevents users from experiencing electrical shocks and may be designed to protect against lightning strikes, thereby providing the same benefits as earthing in any other electrical system. It is necessary to differentiate between different types of earthing/grounding: *protective earthing/grounding (equipotential bonding)* is required to address safety issues; *functional earthing/grounding* is used to improve PV system performance.
- Lightning and surge protection: includes both earthing and surge overcurrent protection against direct and indirect lightning strikes.

Overcurrent protection and disconnection devices

Protection equipment will need to be installed at the string, sub-array and/or array level, on the DC side, as well as on the AC side. When designing the systems protective equipment, the relevant local and international standards specifying the protection sizing guidelines for DC overcurrent protection and disconnection devices should be consulted. Some of these are quoted in Table 4.3.

The voltage rating of the PV array's DC switch-disconnectors depends on the topology of the PV array's components: whether functional earthing is present and whether the inverter is separated or non-separated (transformerless).

Rating the switch-disconnectors to the above requirements ensures that, if a fault occurs, the switch-disconnector will still be capable of isolating the array at the voltage and current experienced.

Table 4.3 Relevant international standards as well as examples of country specific standards related to selecting and sizing overcurrent protection and disconnection devices

	International	Australia	USA	Singapore
String over current protection	IEC/TS/TS 62548–2013 Clause 6.3.4 and 6.3.6.1 and 6.3.7 IEC/TS 60269–6	AS/NZS5033:2014 Clause 3.3.4 and 3.3.5.1	NEC 690.8 (A) (1) and (B)(1a)	CP 5: 1998 Amendment No.1 to CP 5: 1998
String disconnection	IEC/TS 62548–2013 Clause 7.4.1.3	AS/NZS5033: 2014 Clause 4.3.5.2 and 4.4.1.3	NEC690.17	CP 5: 1998 Amendment No.1 to CP 5: 1998
Sub-array over current protection	IEC/TS 62548–2013 Clause 6.3.5 and 6.3.6.2 and 6.3.7	AS/NZS5033: 2014 Clause 3.3.5.2	NEC 690.9(A)	CP 5: 1998 Amendment No.1 to CP 5: 1998
Sub-array disconnection	IEC/TS 62548–2013 Clause 7.4.1.3	AS/NZS5033:2014 Clause 4.2, 4.3.5.2 and 4.4.1.3	NEC690.13 NEC 690.17	CP 5: 1998 Amendment No.1 to CP 5: 1998
Array over current protection	IEC/TS 62548–2013 clause 6.3.6.3 and 6.3.7	AS/NZS5033: 2014 Clause 3.3.5.3	NEC 690.9(A)	CP 5: 1998 Amendment No.1 to CP 5: 1998
Array disconnection	IEC/TS 62548–2013 Clause 7.4.1.3	AS/NZS 5033:2014 Clause 4.2, 4.4.1.3, 4.4.1.4 and 4.4.1.5	NEC690.13 NEC 690.14 NEC690.7(A) NEC 690.17	CP 5: 1998 Amendment No.1 to CP 5: 1998

Table 4.4 Voltage rating for PV array DC switch-disconnectors under different earthing configurations and inverter types as per AS/NZS 5033: 2014

Earthing/grounding configuration	Inverter type	Minimum voltage rating of each pole (conductor)
Equipotential earthing/grounding only	Separated inverter	Overall voltage rating = 0.5 × PV array maximum voltage
Functional earthing/grounding present	Separated inverter	Per pole voltage rating = PV array maximum voltage
Equipotential earthing/grounding only	Non-separated inverter (transformerless)	Per pole voltage rating = PV array maximum voltage
Functional earthing/grounding present	Non-separated inverter (transformerless)	**This configuration is not permitted**

Earthing/grounding

Earthing/grounding arrangements on each site will vary depending on a number of factors including national codes and regulations, installation guidelines for module manufacturers, mounting system requirements, inverter requirements and lightning risk.

There are two types of earthing/grounding that may be required in PV systems:

- Protective (equipotential) earthing/grounding: used to provide protection against exposed conductive parts of the PV system, such as module frames and the mounting structure, from becoming charged. If a person touches a conductive material that is at a higher voltage than the 'ground' they are standing on, they may receive an electric shock. Equipotential bonding is used to keep these exposed conductive parts of the PV system at the same voltage as earth.
- Functional earthing: used in some systems to ensure modules operate properly and achieve expected performance. Functional earthing can be achieved by connecting to the earth either directly or through a high resistance (preferred method). Both methods should have the capability to deal with earth faults.
    - *Direct functional earthing:* an earth fault interrupter (EFI)/residual current device (RCD) should be installed in direct functional earthing. The EFI can interrupt the fault current caused by an earth fault in the system. Direct functional earthing is not the preferred method for functional earthing.
    - *Resistive functional earthing:* a large resistor limits the fault current in an earth fault in resistive functional earthing. The required size of the resistor is dictated by the local standards.

Lightning and surge protection

Lightning and surge protection for a PV array is required in areas that are likely to experience lightning strikes; it consists of a combination of earthing and surge protectors. Earthing protects exposed conductive surfaces from direct lightning strikes and from voltages induced in these surfaces by lightning strikes nearby. Surge protectors protect the conductors and electrical equipment from voltage spikes caused by direct or indirect lightning strikes, by either blocking the surges or shorting them to earth. Lightning protection may also include lightning rods that provide a preferential alternative path to ground for the lightning via earthing cables, reducing the likelihood of the lightning striking the array.

When designing lighting and surge protection it is important to follow all relevant local and international standards. Some of these include:

- IEC/TS 62305 Standard for Lightning Protection
- The 2014 NFPA 780: Standard for the Installation of Lightning Protection Systems, section 12 covers lightning and surge protection for solar arrays.
- AS1768-2007 – Lightning Protection.

AC switchgear

The correct switchgear must be used on the system's AC side to provide disconnection, isolation, earthing, and protection for various components for the site. The type of switchgear required will depend largely on the voltage level reached once it has been stepped up after the inverter.

- For voltages up to 33 kV internal metal clad cubicle type switchgear is often used with gas or air-insulated busbars, and vacuum or SF6 breakers.
- For higher voltages air-insulated outdoor switchgear is the preferred choice, or gas-insulated indoor switchgear if space is an issue.

All AC switchgear should:

- be in accordance with the IEC/TS 61439 series of standards for LV, and the IEC/TS; 62271 series for HV switchgear, as well as relevant local standards;
- have the option to be secured by locks in off/earth positions;
- clearly show the ON and OFF positions with appropriate labels;
- be rated for operational and short-circuit currents;
- be rated for the correct operational voltage;
- in the case of HV switchgear, have remote switching capability; and
- be provided with suitable earthing.

Cable selection and sizing

The selection and sizing of DC and AC cables in a utility-scale PV system is integral to the system's safety, longevity and performance. It is important to design cables in accordance to IEC/TS standards as well as national codes and regulations applicable to the country the solar farm is built it.

Relevant IEC/TS standards include:

- IEC/TS 60364 for LV cabling
- IEC/TS 60502 for cables between 1 kV and 36 kV
- IEC/TS 60840 for cables rated for voltages above 30 kV and up to 150 kV.

Each section of the cabling system should be sized according to the following criteria:

1 The current carrying capacity of the cable: the calculated current should not exceed the safe current carrying capacity (CCC) of the cable. It is important to derate properly taking into account the location of cable, the method of its installation, the number of cores and temperature. The greater the cross-sectional area (CSA) of the cable, the greater the CCC, the more insulated the wire, the less heat is dissipated so the CCC is lower.
2 The cable voltage rating: ensure that the DC cables are rated to carry a voltage greater than the maximum V_{OC} output from the system. The AC cables should be rated to carry the maximum voltage applied at that point.
3 Minimisation of cable losses: the voltage drop between the PV array and inverter, and associated power losses should be kept to a minimum, typically below 3%. The cable routes should be designed to minimise voltage drop and inductive wiring loops. Note that AC voltage rise and power losses between the inverter and the substation should also be kept below the level specified in local standards.

Voltage drop (losses in cables)

In addition to minimising DC voltage drop (losses in DC cables), the AC voltage rise (losses in AC cables) between the inverter and the grid should also be kept to a minimum and are at least less than the maximum permissible figures as specified by relevant standards. Voltage rise occurs because inverters are designed as a current source. The grid voltage then rises as a result of the inverter pushing current onto the grid. A simple way to think of voltage rise is as a negative voltage drop.

In practice, different methods are used to calculate voltage drop (cable losses) in different countries: tables, graphs and mathematical calculations.

The actual CSA of the cable selected will depend on the available cable sizes. For example, typical CSAs for PV module string cables are 1.5 mm², 2.5 mm², 4 mm² and 6 mm². Note that in the US, cable sizes are expressed in terms of American wire gauge (AWG) – see Table 4.5 for conversion between AWG and

Box 4.2 Calculating voltage drop

Voltage drop is calculated using Ohm's law and the resistance of the conductor:

$$V_{DROP} = \frac{2 \times L_{CABLE} \times I \times \rho}{A_{CABLE}}$$

$$\%V_{DROP} = \frac{V_{DROP}}{V_{MAX}} \times 100$$

If the maximum permissible voltage drop is known, the previous equation can be rearranged to calculate the minimum CSA required:

$$A_{CABLE} = \frac{2 \times L_{CABLE} \times I \times \rho}{V_{LOSS} \times V_{MAX}}$$

Where:

L_{CABLE} = route length of cable in metres (multiplying by two adjusts for total circuit wire length).

I = current in amperes.*†

ρ = resistivity of the wire in Ω/m/mm²

A_{CABLE} = CSA of cable in mm².

V_{MAX} = maximum line voltage in volts.‡

V_{LOSS} = maximum permissible voltage loss in the conductor as a percentage, expressed as a decimal, e.g. 3% = 0.03

Note:

*For AC calculations, the current should also account for the power factor:

$$I = I \times \cos\phi$$

†For DC calculations, the I_{MP} current (at STC) should be used and not the I_{SC} current.

‡For DC cables, V_{MAX} is equal to the MPP voltage of the string or array at STC (V_{MP_STRING} or V_{MP_ARRAY}). For AC cables, this is equal to the grid voltage (RMS).

Table 4.5 American wire gauge conversion table

AWG	Size (mm^2)
14	2.00
12	3.31
10	6.68
8	8.37
6	13.30
4	21.15
2	33.62
1	42.41
0	53.50

mm^2. Selecting a cable larger than required will increase the cabling cost but will result in reduced power losses. It is usually cheaper to use larger cables than to install more modules to compensate for the loss.

In addition to surface area, cable route lengths should be minimised in order to keep the voltage drop to acceptable levels, especially for sections carrying high current.

Current carrying capacity ratings for the DC cables

CCC ratings apply to string, sub-array and array cables and should be suitable to the function of the cable and the protection installed on the cable. Descriptions and guidelines of CCC ratings for different parts of the system based on Australian Standards are given in Table 4.6. CCC ratings as specified by other countries' standards will be similar, but need to be confirmed before undertaking the system design.

AC cabling

AC cabling will include:

- Inverter AC cables: these connect the AC side of the inverter to the LV/MV transformer.
- LV/MV transformer cables: these connect the LV/MV transformers to the MV collection switchgear, or substation.
- MV switchgear cables: these connect the MV switchgear to the MV/HV transformers, which connect the system to the grid.

Current national and international standards will apply to AC cabling:

- IEC/TS 60502 for cables between 1 kV and 36 kV
- IEC/TS 60364 for LV cabling (BS 7671 in UK)
- IEC/TS 60840 for cables rated for voltages above 30 kV and up to 150 kV

Table 4.6 CCC rating guidelines, based on Australian standards

Location	Description	Current rating guidelines
String	• Connect PV modules in series • Often terminated in the array combiner box, which connects the array cable to the inverter • Each string cable should be capable of carrying all possible current sources in the system, as well as account for any overcurrent protection limits in the system • The string cables may also carry current being fed from a number of strings to a single string if that single string is not operating at the same voltage level as the other strings.	***If string overcurrent protection will be installed:*** The string cable should be able to carry any current able to pass through the string overcurrent protection. $CCC \geq I_{TRIP}$ ***If string overcurrent protection will not be installed:*** The string cable should be able to carry the combined short-circuit currents from the other strings (with safety margin) as well as any current able to pass through downstream overcurrent protection. $CCC \geq I_n + (1.25 \times I_{SC\ MOD}) \times (N_{strings} - 1)$ **Where** I_{TRIP} = rated trip current of the string I_n = downstream overcurrent protection $N_{strings}$ = total number of parallel connected strings protected by the nearest overcurrent device.
Sub-array	• An array may be broken up into sub-arrays, comprising a number of parallel strings • The sub-array cables connect the string combiner box (the connection point of the parallel strings) and the array combiner box • Each sub-array cable should be capable of carrying the system's total current sources, taking into account any system overcurrent protection • The sub-array cable should be able to carry its short-circuit current • May also need to account for when the sub-array is fed currents from the other sub-arrays because the sub-array is not operating at the same level as the other sub-arrays.	***If sub-array overcurrent protection will be installed:*** The sub-array cable should be able to carry any current able to pass through the sub-array overcurrent protection. $CCC \geq I_{TRIP}$ ***If string overcurrent protection will not be installed:*** The sub-array cable should carry the greater of: 1. Its own short-circuit current (with safety margin) $CCC \geq 1.25 \times I_{SC\ SUB\text{-}ARRAY}$ 2. The combined short-circuit currents from the other sub-arrays (with safety margin), as well as any current that can pass through downstream overcurrent protection: $CCC \geq I_n + 1.25 \times \sum I_{SC_{OTHER\ SUB\text{-}ARRAYS}}$ **Where** I_{TRIP} = rated trip current of the sub-array.
Array	• Connect the PV array to the PV switch-disconnector and then to the DC input of the inverter • Should be capable of carrying all currents from the PV array as well as any possible back-feed current from the inverter • The array cable for a standard grid-connected PV system will not carry any current from external sources, such as a battery bank. Therefore, it is expected that there will be no array overcurrent protection installed.	The array cable should be sized to carry the greater of: 1. The array short-circuit current (with safety margin) $CCC \geq 1.25 \times I_{SC\ ARRAY}$ 2. The inverter back-feed current: $CCC \geq$ inverter backfeed current

- AS/NZS 4777.1:2005 outlines the AC cable requirements in a grid-connected PV system
- AS/NZS 3000: 2009 states the DC and AC derated CCC for different cables
- AS/NZS 4509.2:2009 includes tables outlining the voltage drop per ampere per metre of AC cables for various cable sizes.

Other electrical components

Combiner boxes

In a utility-scale system, string combiner boxes will be installed to house the interconnection of string cables to form a sub-array, along with all necessary protective equipment. In some cases, an array combiner box may be used to house the connection of these sub-arrays before entering the inverter. However, most utility-scale central inverters have multiple DC inputs, so the sub-arrays are paralleled inside the inverter, leaving no need for the array combiner box. This is usually the simplest and cheapest option.

Combiner boxes should be designed to suit the specific requirements of the system (see Figure 4.21). This means it must have enough inputs for the whole sub-array, include appropriately rated fuses and isolators as well as monitoring equipment, and must be large enough to house all of this equipment. It is also important that all combiner boxes located outside have a minimum ingress protection (IP) rating of IP55, are UV stabilised and are high quality to ensure lower losses and to prevent overheating.

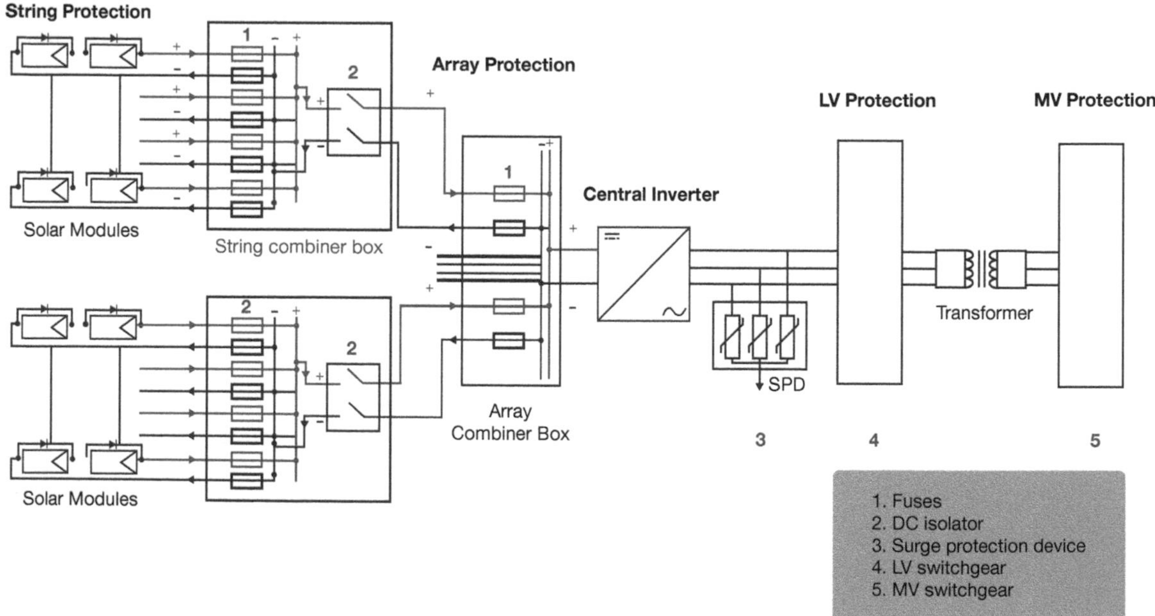

Figure 4.20 Schematic of a portion of a utility-scale PV system, indicating location of protective equipment and combiner boxes.

Specifications		
Number of boxes in question		
Number of strings		
Fuses in	☐ Plus pole (+)	☐ Minus pole (-)
String current [A]		
Max. rated DC voltage [V]		
Total current [A]		
Max. cross-section of the output terminals [mm²]		
Max. cross-section of the string lines [mm²]		
String cable entry	☐ PV WM4 connector	☐ Cable glands
Other cable entry / special features		
Performance monitoring Transclinic xi+	☐ Yes	☐ No
Existing external supply voltage (required for Transclinic xi+)	☐ 24 V DC from outside	☐ 230 V AC from outside
Surge protection (OVP)	☐ Yes	☐ No
Switch-disconnector	☐ Yes	☐ No
Degree of protection (standard: IP55)		
Attachment method	☐ Free-standing (floor distributors)	☐ Wall mounting
Lockable housing	☐ Yes	☐ No
Are special markers needed (e.g. cabinet number in a park)		
Further requirements		

Figure 4.21 This is an example of the options that can be specified to optimise the design for a string combiner box.

Source: Weidmuller

To ensure protection against short circuits, it is recommended that:

- the combiner box enclosure is fabricated from non-conductive material;
- the positive and negative busbars are adequately separated and segregated; and
- the enclosure layout should be such that short circuits during installation and maintenance are extremely unlikely.

System monitoring

Care should be taken when designing the monitoring system for the solar farm. It must be integrated into the electrical design so that it effectively monitors various components of the system. This includes each string of modules, each sub-array, each inverter, each transformer, metering devices, the grid as well as meteorological data such as irradiance, temperature and wind speed. Information from all these components must be sent back to servers in the monitoring station so that it can be accessed remotely via the internet. See Figure 4.22 for an example of a supervisory control and data acquisition (SCADA) system design for a PV plant. For more information on SCADA and system monitoring, see Chapter 6.

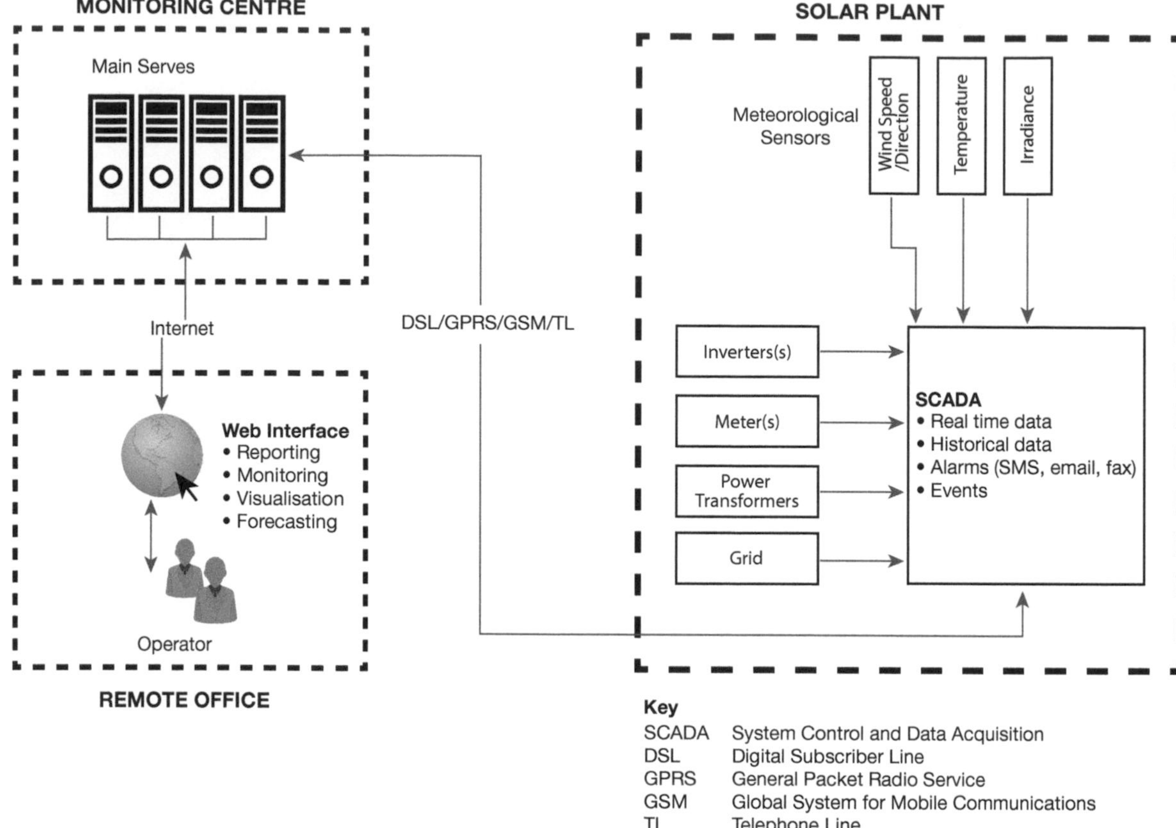

Figure 4.22 PV system monitoring schematic.

Substation design

For utility-scale PV installation, a substation may need to be built to house the electrical equipment for central operation of the solar farm and as well as connection to the local electricity grid. Such equipment includes substation transformers, MV switchgear, monitoring or control systems, protection equipment and metering equipment. The substation can also be used as a base for operation and maintenance staff.

Substation design considerations:

- voltage levels in and out of substation
- optimisation of space
- compliance with relevant building codes and standards
- separation between MV switch rooms, converter rooms, control rooms, store rooms and offices
- safe access, lighting and welfare facilities
- working space for O&M staff

- HV safety – earthing/grounding mat may need to be installed prior to setting the foundation to provide safe step/touch potentials and earth/ ground system faults
- metering – tariff metering will be required to measure the export of power and may be installed at the substation
- system monitoring – provide control and status indication for components in the substation and across the solar plant; air conditioning should be considered due to the heat generated by the electronic equipment in the modules
- auxiliary equipment required for a functioning substation/control room.

There is also the option of hiring a third party to undertake the design and installation on-site of the substation to cut time and costs. In the US, companies like ABB and S&C Electric are delivering pre-engineered, pre-packaged substations that can be tailored to the specific project requirements.

System design and drawings

Once the electrical design is completed and approved and all the primary system components have been selected, a detailed system design should be drawn up including:

- Architectural diagram: a plan view of the installation site drawn to scale. It will show the physical dimensions of the site, the location of all relevant existing equipment and the proposed location of all system equipment. This should be done using computer software such as AutoCAD.
- List of all materials required: this represents the installation system equipment checklist.
- Electrical schematic: also known as a 'circuit diagram' and; a drawing of the circuitry of systems, components, devices, represented by graphical symbols and their interconnections. The location of components in the diagram do not necessarily correspond to the actual physical arrangement of the equipment, but provides an understanding of the electrical system.
- Wiring diagram: also known as a 'connection diagram'; a detailed drawing showing the wiring and interconnectivity of all components and their actual locations, e.g. the array combiner box or the AC wiring in the switchboard. The wiring diagram may include the function of the components, the wire type, size and colour, and the identification of the circuits leading to and from the component. These diagrams are used when assembling, installing or maintaining equipment.
- Block diagram: a drawing used to give an overview of the system, including the function of the system (a 'single-line diagram' is a type of block diagram). There may be no need to have a block or a single-line diagram if all the information can be included in the circuit diagram.

Computer-aided design optimising

Computer-aided design optimisation tools are very useful during all stages of the design process described in this chapter, as well as in the preliminary stages, when submitting project proposals or seeking finance. These tools allow for easy simulation of multiple design configuration options, ensuring the optimal solution can be realised.

Once the design is finalised it will need to be drawn up accurately using computer-aided design (CAD) and/or geographic information system (GIS) software. CAD software will assist a designer in the creation and integration of electrical and mechanical components whilst GIS software will allow the designer to integrate the design with real world data such as site topography, stormwater drainage, electrical distribution or transmission infrastructure location, real world grid geo-reference points, etc. As an example, the software company Autodesk provides a software suite that a designer might use to design a utility-scale solar farm. Autodesk Map 3D might be used for the GIS design and Autodesk AutoCAD civil 3D might be used for the CAD component design.

There are many software packages available on the market which are able to produce similar outputs to the ones shown in Figure 4.23. Many companies have produced software which is specific to the PV industry with specific tools, features and functionality which can assist the designer to be more accurate and efficient. Further, there are many plugins or extensions which can be added to the core CAD/GIS package produced by some of the more well-known software companies. This allows designers to utilise a common platform across design tasks.

The CAD/GIS designs are incredibly important to the success of the project. If a highly accurate PV system design is created with a high level of quality assurance, the project will have a much lower risk profile and will ultimately cost much less. To provide an example of the importance of CAD/GIS design,

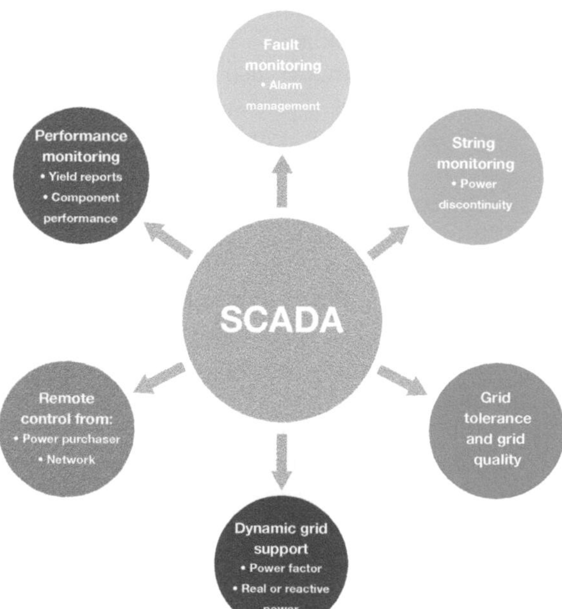

Figure 4.23 Terrain data can be analysed using HELIOS 3D, marking which parts of the site have north, east, south or west orientation.

Source: Stoer+Sauer

Figure 4.24 Solar farm design and layout using HELIOS 3D.

Source: Stoer+Sauer

much of the new pile-driving equipment on the market uses GPS positioning equipment, which can be pre-loaded with the geo-referenced PV system design. An accurate design ensures the accurate location and height of structural members with minimal on-site effort and low risk of operator error.

HELIOS 3D is a tool designed specifically for utility-scale solar systems and uses AutoCAD Civil 3D to place 3D solar racking in a 3D space. This tool has been referenced throughout the planning and design chapter as it has a wide range of functions and can be used for many stages of the project including:

- Project development: analysis and evaluation of the terrain, rate of yield.
- Project layout: structuring of the terrain, positioning of PV racks, optimise positioning for maximum yield.
- Project engineering: electric layout, trenching, cabling, and string assignments
- Evaluation and documentation: bill of materials, list of GPS coordinates, cable lists.

PVSyst is another available tool that can be used in conjunction with HELIOS 3D to accurately analyse different design configurations of remote or grid-connected PV systems. It evaluates the results and identifies the optimal design solution, to provide the highest system performance and revenue. An analysis will be provided of the system's behaviour based on irradiation, temperature and shading information, and suggests potential improvements to the design. The product comes with a complete database of PV panels, inverters and meteorological data, as well as the following features:

- Import of irradiation data from PVGIS, NASA databases.
- Import of PV modules data from PHOTON INTERNATIONAL.
- Useful 3D application to simulate far and near shadings.
- Specialised tools for the evaluation of the wiring losses (and other losses like the module quality), the mismatch between modules, soiling, thermal behaviour according to the mechanical mounting, system unavailability, etc.
- Calculation of total energy production (MWh/year), PR and specific yield (kWh/kWp).

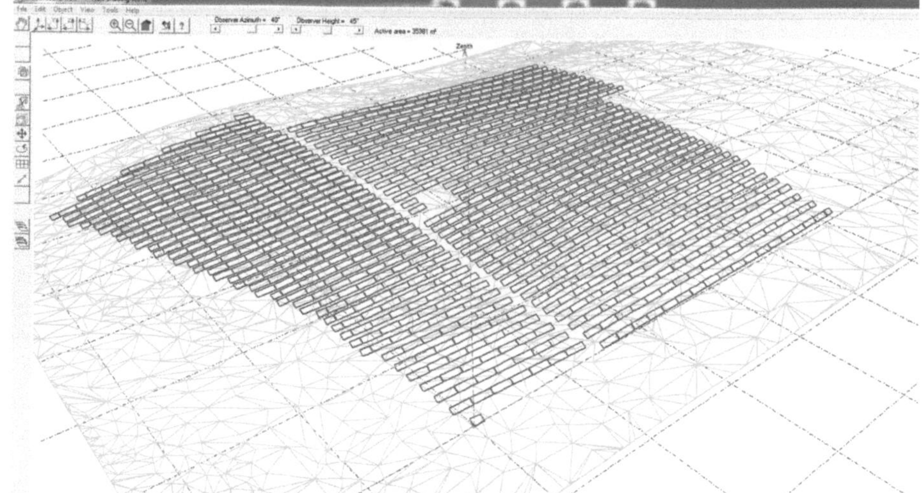

Figure 4.25 A complete 3D layout of a project generated by HELIOS 3D can be exported quickly and imported in PV-Syst.

Source: Stoer+Sauer

- Economic evaluation and payback.
- Export of calculations to CSV files.

System equipment

Solar module selection

The choice of module type for any solar installation can be influenced by a number of factors, e.g. by an existing business relationship or a government preferential requirement. However, the cost and performance of different types of modules for the given site-specific conditions should be calculated and compared. The choice of the most suitable type of module is not always easily made. Although module manufacturers may specify the power rating and performance of the module under specific irradiance, temperature and voltage conditions, the performance may vary significantly under different conditions. Typical PV module characteristics are shown in Table 4.7 and consideration of any or all of these may be applied to the selection of solar module used.

It is also important to consider the quality benchmarks of the modules, which include:

- Product guarantee: manufacturers ensure the product will be fully functional for a specified period (typically two to six years).
- Power guarantee: manufacturers ensure the module's operating performance for a certain period of time. Typical values are 90% of the original nominal power after ten years and 80% after 25 years.
- Lifetime: the desired lifetime estimate of a solar module should be 25–30 years.

Table 4.7 PV module characteristics and considerations

Characteristic	Considerations	Module types
Efficiency	• Modules quoted at a high conversion efficiency are more expensive, but, for a given installed capacity, will require significantly less land, cabling and support structures • Modules having lower conversion efficiency may be suitable when the amount of available installation space is not an issue	• **Monocrystalline** modules have the highest efficiency but are the most expensive • **Polycrystalline** have a lower efficiency than monocrystalline but higher than thin film • **Thin-film** modules are the cheapest but least efficient
Temperature coefficient	• For hot climates, it is important to choose modules with low temperature coefficients to reduce efficiency losses due to high temperatures	• **Thin-film** modules have the lowest temperature coefficient: meaning that the efficiency of thin-film modules is less affected by high temperatures compared to crystalline modules, and thus, in some circumstances, may make them a preferable option for very hot locations
Number of cells	• 60-cell modules are more commonly available and have lower per module cost but higher installation costs • 72-cell modules are larger than 60-cell modules and therefore have a higher cost per module, but these larger modules will require less mounting hardware and labour, therefore the installation costs are lower • Note that 48- and 96-cell modules also exist	• This only applies to **crystalline** modules
Shade tolerance	• If there is unavoidable shading at the site, the system's performance may be increased by selecting a module with good performance in low light conditions	• **Thin-film** modules have a better shade tolerance than crystalline modules

These benchmarks can be achieved if the modules have been manufactured to a high quality. It is therefore very important to select a high-quality and reliable module manufacturer and supplier. Some indicators for selecting manufacturers are given in Table 4.8.

Inverter selection

The inverter is the heart of a PV system. For utility-scale systems, central inverters are most commonly used, although string inverters and even micro-

Table 4.8 Signs of manufacturer quality

Quality indicators	Description
Process integration	Vertically integrated companies, in which each step of the manufacturing process is done by one company, are generally mature PV manufacturing companies with a commitment to the industry
Quality control	Companies with strong commitment to quality and automation of manufacturing are likely to consistently produce high-quality solar modules
R&D	Companies that invest in R&D are generally committed to the industry and will be more likely to keep up with industry changes
Volume and length of time of solar module production	Companies that produce in high volume and/or rapidly are likely to be larger and have more experience
Market representation	Companies that have multiple products across several world markets may have a better reputation
Insurability	If a planned large-scale system cannot get insurance, the project is unlikely to go ahead. Therefore, designers might consult insurance companies to see which product manufacturers they 'trust' based on various factors such as independent testing and past experience
Bankability	The ability of a manufacturer's product to attract investment in the past provides historical context to help judge the reputation of the manufacturer. Companies can therefore be ranked on their bankability (e.g. Bloomberg New Energy Finance's PV Module Maker Tiering System), although this is not a substitute for due-diligence checks, as it is not a direct measure of quality

inverters may also be used. Selecting the inverter to be used is a critical design decision as this influences the project costs, installation methods and electrical configuration. Table 4.9 sets out a list of advantages and disadvantages for each type of inverter. More information on different types of inverters is given in Chapter 2.

In addition to the type of inverter, there are many different capacities and functions offered by inverters and all these features must be considered when selecting the most suitable inverter. Some of these considerations are set out below.

- Incentive schemes
  - The combined banding of financial incentive mechanisms may have an influence on the choice of inverter.
  - For example, applicable feed-in tariff (FiT) schemes might be tiered according to the installed capacity of the solar farm and this may, in turn, influence the inverter size.
- Project size
  - The project's size influences how to configure the interconnection of the inverter to the network. Central inverters are most commonly used in utility-scale solar applications.
  - Different inverters have different voltage, current and power specifications. Different specifications will be better suited to different projects and array configurations.

Table 4.9 Comparison of different types of inverters. Note that having high redundancy means that less of the system is affected if an inverter fails

Inverter type	Advantages	Disadvantages
Micro-inverters (200 W–300 W)	• Increased system yield for arrays that have modules that are not subject to the same operating conditions • High levels of redundancy • Reduced DC cabling and protection requirements	• Large number of units increases system cost • Increased array weight
Single-phase string inverters (1 kW–5 kW)	• Cheaper inverter units • Lower capacity means flexibility in total system inverter capacity • High levels of redundancy	• Large number of units increases system cost • Increased Balance of System (BoS) equipment related to having to connect large number of units • Basic inverter unit with minimal output control and monitoring
Three-phase string inverters with module MPPTs (5 kW–60 kW)	• Increased system yield for arrays not having constant operating conditions • High levels of redundancy	• Large number of units increases system cost • PV array/inverter layout needs to be configured so that output is balanced across the three phases
Three-phase string inverters (5 kW–25 kW)	• Larger capacity means fewer inverter units • Flexibility in total system inverter capacity • Array is balanced across three phases • Increased monitoring and output control capabilities • High levels of redundancy	• Reduced system yield for arrays with varying operating conditions
Central inverters (50 kW–2200 kW)	• Largest capacity means minimum number of inverter units • Advanced monitoring and output control capabilities	• Reduced system yield for arrays with varying operating conditions

- Performance
  - High-efficiency inverters should be chosen.
  - The additional yield produced by a higher-efficiency inverter usually more than compensates for the higher initial cost.
  - Consideration should be given to the fact that efficiency changes with DC input voltage, percentage of load, and several other factors.
- MPP range
  - A wide MPP range allows flexibility and facilitates design.
- Three-phase or single phase output
  - National electrical regulations might set limits on the maximum power difference between the phases in the case of an asymmetrical load.
  - For example, according to the German grid code the maximum difference allowed between phases is 4.6 kVA. Therefore, a three-phase system proposed for installation on a German grid would have to be carefully modelled to quantify the difference in power supplied across the phases.

- Module technology
  - The compatibility of thin-film modules with transformerless inverters should be confirmed with manufacturers.
- National and international regulations
  - A transformer inverter must be used if galvanic isolation is required between the DC and AC sides of the inverter.
- Grid code
  - The prevailing grid code affects inverter sizing and technology.
  - The national grid code might require the inverters to be capable of reactive power control. In that case, oversizing inverters slightly could be required.
  - The grid code also sets requirements on total harmonic distortion, which is the level of harmonic content allowed in the inverter's AC power output.
- Product reliability
  - High inverter reliability ensures low downtime and maintenance and repair costs. If available, inverter mean time between failures (MTBF) figures and the product's track record should be assessed.
- Module supply
  - If modules of different specifications are to be used, then string or multi-string inverters are recommended, in order to minimise any performance losses because of module mismatch.
- Maintainability and serviceability
  - Ease of access to qualified service and maintenance personnel, and availability of parts is an important dimension to consider during inverter selection. This may favour string inverters in certain locations.
- System redundancy
  - If a fault occurs with a string inverter, the loss represents only a small proportion of the plant's output. Replacement inverters could be kept locally and replaced by a suitably trained technician or electrician.
  - If a central inverter is out of commission, this usually represents a large proportion of the plant output that would be lost (for example, 100 kW) until a replacement is obtained.
- Modularity
  - Ease of expanding the system capacity and flexibility of design should be considered when selecting inverters.
- Shading conditions
  - For sites with different shading conditions or orientations, string or micro-inverters might be more suitable.
- Installation location
  - Whether the inverter is to be installed outdoors or indoors and the site's ambient conditions will influence the IP class and cooling requirements of the inverter to be used.
- Monitoring/recording/telemetry
  - Plant monitoring, data logging, and remote control requirements represent a set of criteria that must be taken into account when choosing an inverter.

Transformers and harmonic filters

As discussed in Chapter 2, there are two different types of transformers typically used in a utility-scale system: distribution and substation transformers. When specifying or selecting the most appropriate transformer, it is important to consider the required size of the transformer (see transformer sizing, under the electrical design section), its position in the electrical system (Figure 4.26) and the physical location of the installation.

Transformers are often customised for the proposed system design. Each solar farm will require different specifications: this will include important device ratings such as:

- voltage: critical to specify correct nominal voltage level;
- winding connections: the primary and secondary winding connections;
- basic impulse level: the level of momentary overvoltage that the transformer insulation can withstand without damage or failure;
- impedance: the amount of resistance and reactance in the circuit;
- efficiency: a measure of output power relative to input power (higher for larger transformers);
- winding material: copper offers better electrical conductivity by volume than aluminium, while aluminium offers better electrical conductivity by mass than copper;
- temperature rise: defined as the average temperature rise of the windings as compared to the ambient temperature (higher temperature rises mean higher efficiency losses);
- insulation class: higher insulation class means lower temperature rise;

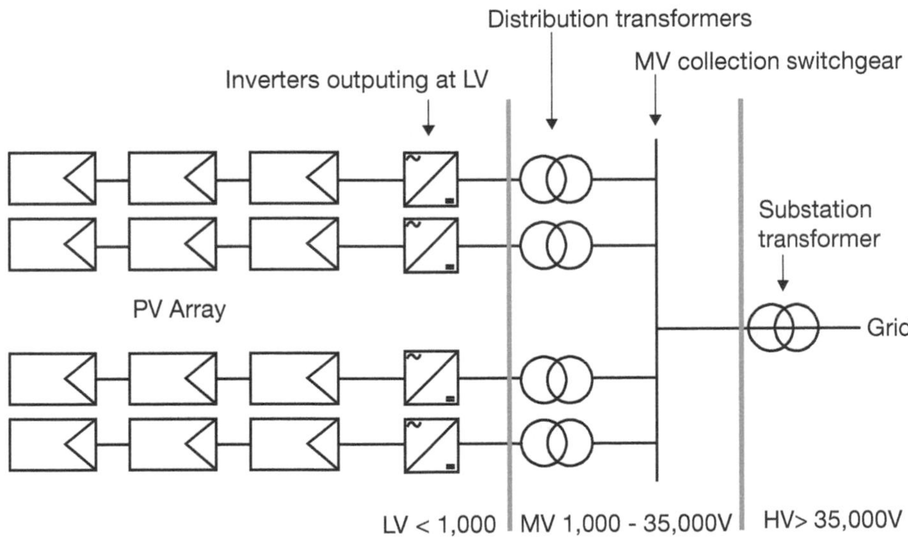

Figure 4.26 Typical transformer locations and voltage levels in a utility-scale solar plant.

- overcurrent protection: in the event of a fault, an expulsion fuse is used to limit its duration and a current-limiting fuse limits its magnitude;
- overvoltage protection: during a voltage surge on the grid, overvoltage protection devices are used to quickly limit the overvoltage by conducting the surge current to ground; and
- switch-disconnects: used to isolate individual transformers from the primary circuit.

Harmonics and harmonic filters

Inverters produce harmonics of varying degrees resulting in harmonic currents flowing through the transformer. This can lead to overvoltage, energy losses, and further damage to the transformer. It is important to specify the harmonic loading on the transformers so they can withstand these harmonics and the overheating caused by the harmonics, this can be achieved through thermally oversizing the transformer.

Harmonic filters can also be used to reduce the harmonic impedance at the resonance frequency and therefore reduce the potential damage to transformers. Three different types of harmonic filters can be used depending on the cause of harmonic distortion: these include passive filters, active filters and hybrid filters. The passive filter is most commonly used in large-scale PV systems and can be integrated into the inverter design.

Mounting systems

The mounting system chosen for the PV array in a solar farm should be chosen to produce maximum yield, should be cost-effective, and should provide the appropriate structural support to withstand the real-world environment. Some important considerations are explained in Table 4.10.

A comparison between fixed and tracking mounting systems is given in Figure 4.27. Although fixed mounting systems are simpler and cheaper, if the higher costs of tracking systems are offset by an increased yield, then a tracking system may be the most suitable choice.

Single-axis trackers move the solar modules in one (east–west) axis to follow the sun's path from dawn to dusk (see Figure 4.28). The increase in output power from these systems compared to 'fixed' mounting systems is illustrated in Figure 4.29.

A dual-axis tracker follows the sun's path from dawn to dusk (i.e. east–west) and adjusts for the seasonal movement of the sun from summer solstice to winter solstice, by changing the tilt of the modules to be flatter during summer and steeper in winter (Figure 4.30). Dual-axis trackers are most effective in locations further away from the equator, where the seasonal variations in the position of the sun are most pronounced.

Table 4.10 Issues to be taken into consideration when selecting mounting structures

Issue	Why is this is an issue?	Possible solutions
Wind loading	• Wind can exert downward, upward, and bilateral forces on the mounting structure. These forces can be exerted separately or in combination. • Mounting systems must be designed and installed so that all mounting structures can withstand the site-specific structural loadings caused by the wind. • Incorrect design and/or installation can lead to the system modules being damaged or blown away by high winds.	• Earth-screws or helical piles may be the most suitable foundation option if there are high wind speeds, because they provide high pull-out resistance.
Ground type and soil conditions	• Different mounting structures are suited to different soil types and topographies. • The results from the geotechnical and topographic survey should be used to determine the suitability of different mounting structures. • Weather data is required which provides rainfall data for the proposed installation site. The soil type, topography and proposed system layout must be confirmed as suitable in relation to the site-specific weather data.	• Trackers require flatter sites than fixed mounting systems (generally < 5° slope). • If soil is prone to erosion, concrete ballasts may be needed to provide extra support. • If the site is prone to flooding, mounting structures will need to be suitably elevated.
Available area	• The available area at the site must be confirmed as suitable to accommodate the proposed system configuration.	• Depending on the shape and dimensions of the available land, more modules may be able to be installed based on using landscape or portrait orientation. • If the available area is limited, tracking systems may be the best option to deliver the proposed output from the solar farm.
Row height	• If rows are multiple modules high, the mounting structure must have suitable foundations to support the increased weight and wind loading. • Installing modules above other modules on each structure increases the complexity and cost of installation, but this means that fewer rows are needed for the array. • There must be enough space between rows: ○ To allow time and cost-efficient system installation ○ To minimise inter-row shading ○ To accommodate maintenance vehicles.	• Rows may be one module or multiple modules high.

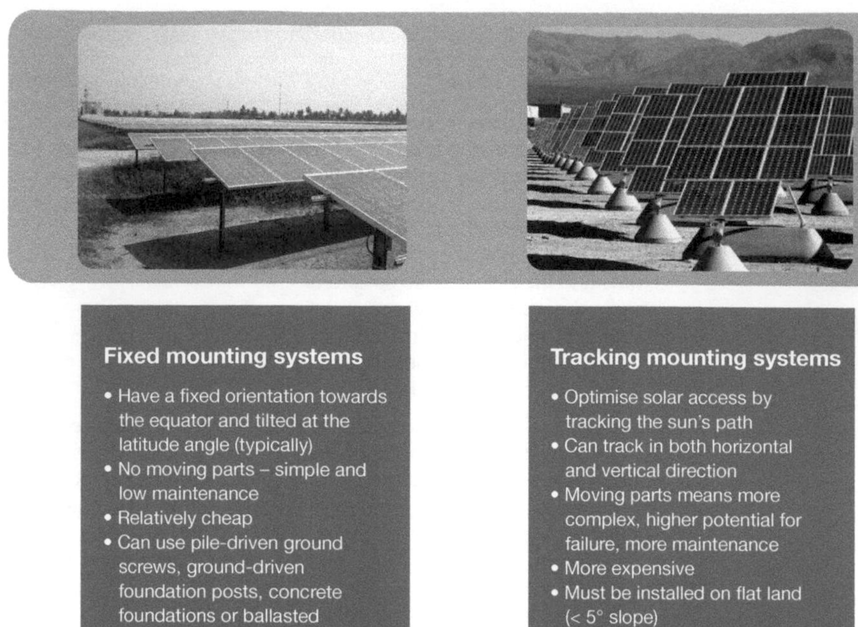

Figure 4.27 Comparison of fixed and tracking mounting systems.

System integration

A utility-scale solar system needs to be designed so that it can be smoothly integrated into the local electricity grid. This will require close communication with the local utility to ensure that the design is compliant with their guidelines.

Before any generator can connect to a power network:

- The voltage level of generation should match the voltage level of the external network.
- The frequency of generation should match the frequency of the external network.
- The point of cycle (or phase angle) for each phase of generation should equal the point of cycle for each corresponding phase of the external network.
- Compliance with the network's specific requirement for protective devices: this range of protective devices will have specified features, e.g. to operate as specific relay or switch. These requirements can vary across networks and countries. These device numbers are stated in ANSI/IEEE C37.2–2008 Standard Electrical Power System Device Function Numbers, Acronyms and Contact Designations.

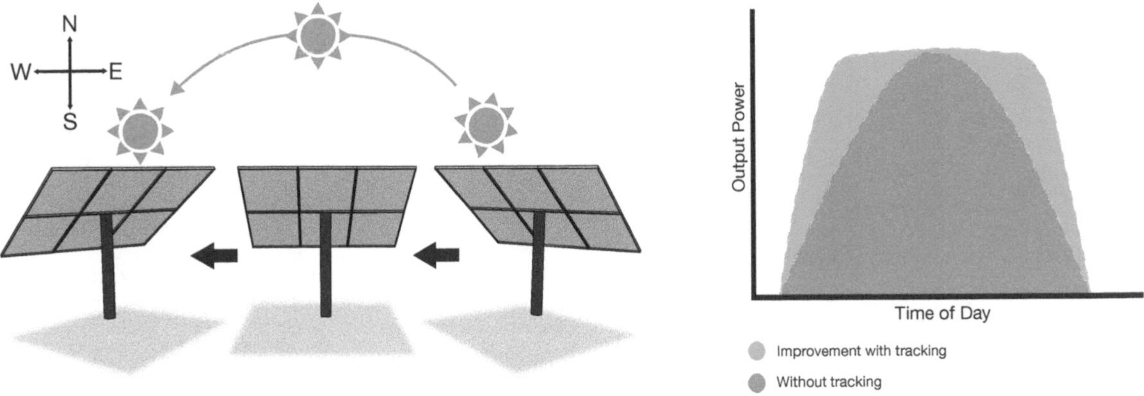

Figure 4.28 A single-axis tracker rotates the modules from east to west during the day to face the sun as it moves across the sky.

Source: Global Sustainable Energy Solutions

Figure 4.29 Benefit of horizontal-axis tracking systems.

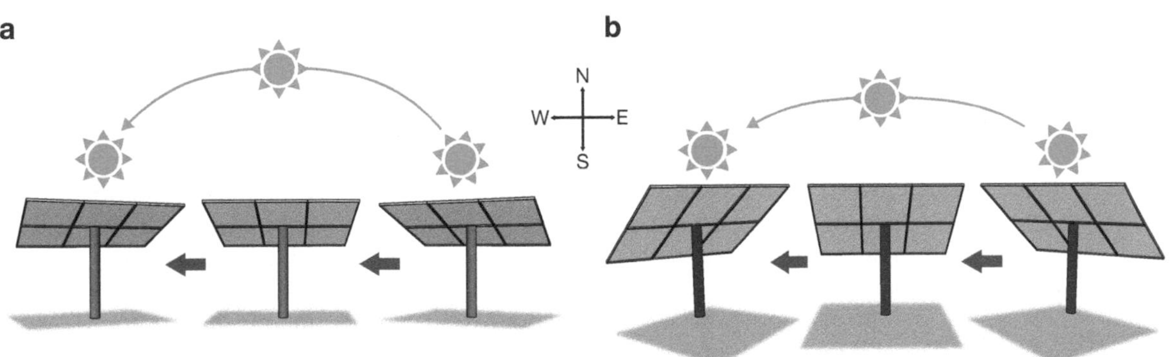

Figure 4.30 A dual-axis tracker moves from east to west during the day and adjusts the tilt as required throughout the year. The modules will be flatter in summer (a) and steeper in winter (b).

Source: Global Sustainable Energy Solutions

The solar farm design should include a SCADA and control system that is capable of controlling the quality of the output power from the solar farm by monitoring the actual grid conditions, and sending instructions to each inverter based on location, losses and performance.

For more information on grid integration and interconnection, see later section on meeting grid-connection requirements.

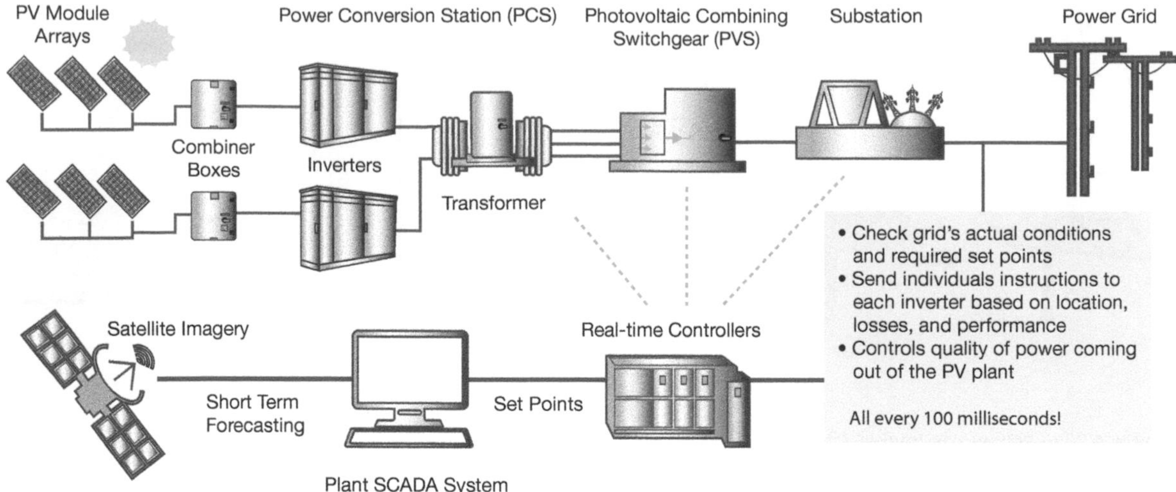

Figure 4.31 Grid integration and plant control system.

Foundations

The choice of foundation type will be based on the topographic conditions of the site and information gathered from the geotechnical survey conducted during the planning stage as discussed in Chapter 3. The foundation type chosen will then govern the type of support system design installed. A suitable foundation system should be easy to install, withstand the weather conditions at the site, and be made from high-quality, corrosion-resistant material. An overview of the suitability of different foundation options is given in Table 4.11. For more detailed information, see the section on foundation construction in Chapter 5.

However, this is just a rough guideline, choosing the right foundation requires consideration of a range of environmental and engineering factors including:

- Soil conditions
  - consolidation
  - permeability and seepage
  - bearing capacity
  - lateral earth pressure
  - slope stability
  - contamination.
- Wind loads
  - wind speed
  - module size and weight
  - orientation.

To summarise:

- Earth screws may be suitable for sites with soft soils, high wind speeds and rocky terrain, but may be complex and expensive to install.

Table 4.11 Comparison of different foundation types according to their attributes

Foundation type	Suitable for soft soil (low pull-out resistance)	Suitable for rocky terrain (high refusal)	Suitable for sloped/uneven terrain	Require few specialist skills and machinery	Low invasiveness	Cheap and easy installation
Driven piles	×	×	✓	✓	✓	✓✓
Earth screws	✓	✓	✓	×	✓	×
Helical piles	✓	×	✓	×	✓	✓
Ballasts	✓✓	✓✓	×	✓✓	×	×

- Helical piles are suitable for soft soils, high wind speeds and are easy to install but not suitable for rocky terrain.
- Driven piles are cheap and easy to install, but have low pull-out resistance so are more suited to cohesive soils like clay and dense sand, may not be suitable in areas of high wind speed.
- Ballasted systems are suitable for soils that are difficult to penetrate, are too soft to hold piles, are rocky or if there are subsurface contaminants; however, they are only suitable for flat terrain.

Grid connection

Grid connection involves both the physical connection arrangement including the construction of electrical infrastructure as well as ensuring that the grid-connection requirements are met to avoid network instability. Grid-connection requirements will vary depending on specific site conditions and local regulatory structure, hence it is important to carry out a thorough investigation into these requirements through early engagement with network operators.

Physical connection arrangement

The physical connection arrangement will depend on existing network infrastructure including the connection capacity available and the proximity to the existing grid-connection point. Physical connection to the grid may include:

- connection to and/or extension of current transmission lines;
- upgrading existing transmission lines;
- construction of new transmission lines; and
- construction of substation or switching station within the boundaries of the solar farm.

The costs and timescales of connection options must be considered as well as negotiation with network service providers to determine who is responsible for capital costs and operation and maintenance.

PV generation intermittency

Although solar irradiance can be predicted to follow seasonal and daily patterns, local weather can present unpredictable weather variations including cloud cover. The variability that unexpected PV generation intermittency produces can result in the local electricity network being negatively impacted: the impact can be local, e.g. a feeder or substation, or broader in effect, meaning it affects the network.

Where the network is unable to support the intermittency of the PV power by base load generation, there can be problems with the network's power quality and potentially network failures. The potential installed capacity of the proposed PV solar farm may be limited because of these possible network problems.

The PV intermittency generation can produce problems over different timescales:

- Changes in power quality: fast changes in PV generation may cause power quality problems, e.g. flickering, harmonic distortion. The network should be able to respond to this in seconds.
- Network's total capacity: the network's power generation must be able to meet the capacity and ramping requirements if there are short-term variations in demand and PV generation. This can result in issues of power quality and system outage. The network will respond to this in minutes of the problem.
- The network's ability to the generating unit's commitment and scheduling: the network's generation planning requires that there is enough generation at any point in time to meet the demand and the operating reserve. The network will respond to this problem within hours to days.

The network demand must be well understood before determining the level of PV able to be successfully interconnected.

Figure 4.32 provides an example of clear sky irradiance and the changes to that irradiance caused by intermittent cloud cover at that site.

The system reliability calculations may indicate that it is preferable to install the required PV capacity over a larger area and a larger number of sites, assuming the network interconnections are viable, as this can reduce the performance variability.

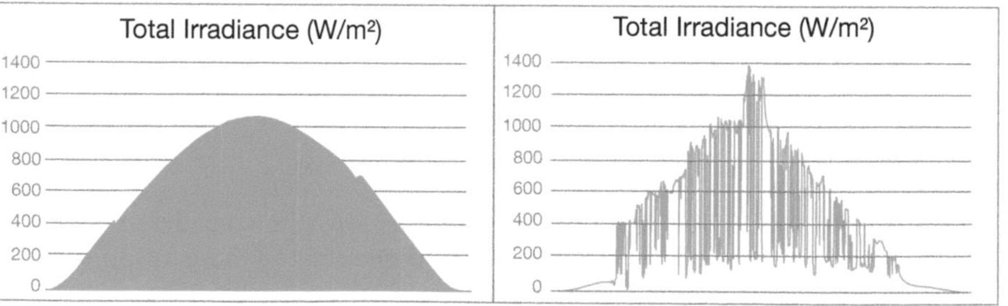

Figure 4.32 Comparison between irradiance on a day with a clear sky, and irradiance on a day with intermittent cloud cover indicates the lack of reliability that intermittent PV generation introduces to any network.

Source: CAT Projects & ARENA

Meeting grid-connection requirements

Since the majority of electricity networks around the world have been designed for electricity flowing from large traditional centralised generators such as coal-fired power stations, they have not been designed to handle the intermittency of large-scale solar plants and this can lead to significant grid integration challenges as described above. In order to ensure that the solar farm will not have adverse impacts on the network stability or system security, thorough grid-connection studies must be carried out in accordance with the local grid-connection rules and requirements. International guidelines include IEEE 1547 and IEC/TS 61727. Guidelines for interconnection procedures for generating facilities in the US have been published by the Interstate Renewable Energy Council (IREC) (available at www.irecusa.org/regulatory-reform/interconnection).

The intermittency of solar generation can lead to changes in the local network's frequency and voltage, which can cause serious challenges for network stability, power quality and security of supply. Different networks have different voltage and frequency requirements which vary between countries, states, regions, etc. so it is very important to remain within the local grid's voltage and frequency tolerance.

A suitable power factor, reactive power and voltage control strategy should be developed to ensure that the power quality of the solar farm remains within utility requirements. One solution is to install an active control system that reads the output of the inverters and compare it to the power factor required by the local grid. If the power factor is too low, the control system calculates a control signal to adjust the output of the inverter. A power factor range of 0.95 lagging or leading is typical for most networks.

In general, the solar farm control system should include the following in order to effectively manage grid reliability and stability:

- reactive power capability to regulate the power factor and plant voltage/ VAR controls;
- active power regulation to curtail active power when necessary;
- ramp rate control to limit the ramp rate from variation in irradiance;
- ride through capability to prevent faults and other disturbances; and
- frequency droop control to monitor, track and react to changes in grid frequency.

Overall, the development of an adequate grid-connection strategy should be done early, and in close consortium with local network service providers to avoid costly time delays, and to ensure smooth integration of the system to the local grid. For more information on power output controls for grid stability, see Chapter 6.

Yield and loss analysis

The analysis for a system's projected yield and loss figures is crucial to be completed in the primary design phase, and is used to predict the annual average output of the solar farm throughout its lifetime after having estimated and accounted for all the losses attributable to the operating conditions and the system equipment used. Yield and losses analyses consist of:

- detailed solar resource and shading analysis during site assessment;
- loss factor analysis and calculation of performance ratio (PR);
- annual yield;
- annual degradation estimation; and
- variation and sensitivity analysis.

Solar access and shading analysis

The solar access and shading analysis is used to calculate the site-specific irradiation which is the starting point for the calculation of the performance of the system. This analysis will be based on the solar resource assessment conducted during the planning stage, outlined in Chapter 3, however it will be more detailed and take the specific design parameters into account.

Since data measured will be global horizontal data, it is critical to account for the module tilt angle and orientation when determining the solar access for the array. There is a range of software available to convert global horizontal irradiance to global tilted irradiance, based on the tilt of the modules at the site. These include:

- PVGIS (Figure 4.33) http://re.jrc.ec.europa.eu/pvgis/index.htm
- Homer Energy www.homerenergy.com/software.html
- PVsyst www.pvsyst.com/en
- US National Renewable Energy Laboratory www.nrel.gov/rredc
- Gaisma: www.gaisma.com
- NASA website provides both horizontal and in-plane data https://eosweb.larc.nasa.gov/cgi-bin/sse/global.cgi
- Commission of the European Communities (CEC) provides tables to convert from horizontal to in-plane data.

If a tracker is used, solar access throughout the year will be increased. Additionally, it is important to break up radiation data into direct and diffuse components: the higher the ratio of direct to diffuse radiation, the higher the yield of the solar farm. This is particularly true for solar farms using tracking systems, because the tracker follows the sun's direct beam.

A detailed shading analysis should be conducted to account for any unavoidable shading that may occur on any part of the array at any time of year. This shading analysis can be developed by a three-dimensional model of the array and all its surroundings. This model could be coupled with half-hourly irradiation data

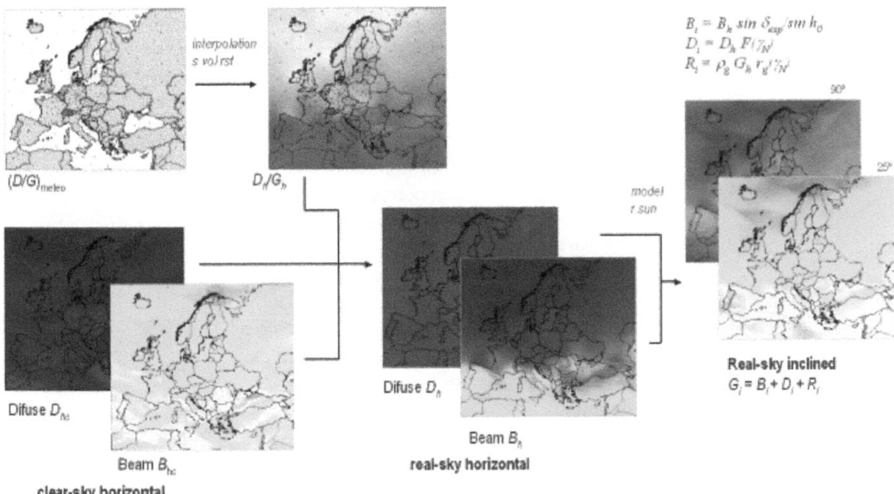

Figure 4.33 PVGIS is a tool used to estimate beam, diffuse and reflected components of the clear-sky and real-sky global irradiance/irradiation on horizontal or inclined surfaces. This image shows the calculation scheme for irradiation at an inclined surface.

Source: PVGIS

using computer software such as PV-Syst and HELIOS 3D, as discussed previously to determine the extent of any shading that may occur on-site.

Shading is variable and can affect the output of the array differently depending on the configuration of the solar modules (i.e. number of bypass diodes and MPPTs) and how the inverter responds to lower voltage output due to shading. It is therefore important to determine the worst-case scenario for the identified shading and to clearly document all assumptions made in the shading analysis.

Energy losses and performance ratio (PR)

The performance of the solar farm can be quantified by the term: performance ratio (PR). PR is discussed in Chapter 3, in relation to operational risks. During the planning stage, the developer must guarantee a certain PR in order to secure finance, and will be liable if the performance of the plant falls below this. The PR measures the quality and reliability of the plant irrespective of location and solar insolation. It is therefore very dependent on how the system is designed. During the design stage, the system designer will be accountable for the PR of the solar farm, and must aim to design the system so that it performs to the guaranteed PR throughout the project lifetime.

When designing the system, the PR should be estimated based on the design parameters and assessable loss factors. There is a large range of energy losses that impact the PR of a solar farm which should be taken into consideration when designing the system. Some of these losses are described in Table 4.12.

Table 4.12 Typical losses in a PV power plant used to calculate the PR

Loss	Description	Approximate loss factor
Module temperature (f_{temp})	The characteristics of a PV module are determined at standard temperature conditions of 25°C. As discussed previously in electrical design, for every degree rise in Celsius temperature above this standard, modules reduce in efficiency. In high ambient temperatures under strong irradiance, module temperatures can rise significantly, reducing the power output. Different types of modules will have different power temperature coefficients.	10–20%
Soiling (f_{dirt})	Losses due to soiling (dust, bird droppings, debris etc.) can be highly variable depending on the environmental conditions (how dusty, salty or polluted the air is), rainfall frequency, tilt angle (higher tilt angle means it is easier for rain to wash away dirt or debris), and on the cleaning strategy as defined in the O&M contract.	5–15%
Manufacturer's tolerance on module quality (f_{man})	Most PV modules do not exactly match the manufacturer's nominal specifications due to errors in measuring equipment and module mismatch (when modules in a string do not all present exactly the same current/voltage profiles due to statistical variation in manufacture). The loss due to manufacturer's tolerance on module quality quantifies the loss of energy yield due to divergences in actual module characteristics from the specifications.	0–3%
Voltage drop in cables ($f_{V_{drop}}$)	Electrical resistance in the cable between the modules and the input terminals of the inverter (DC) and from the inverter to the substation (AC) give rise to voltage drop, and hence power loss. This loss increases with temperature and cable length, and reduced surface area. If the cable is correctly sized, this loss should be less than 3% for the DC side and 1% on the AC side.	< 3% (DC) from furthest module to inverter e.g. AS 5033 <1% (AC)
Inverter performance (f_{inv})	The inverter converts DC into AC with an efficiency that varies according to the inverter load. Losses are in the form or heat, and therefore the inverter efficiency is directly proportional to operating temperature.	2–6%
Transformer performance (f_{trans})	The transformer's performance is measure by how efficiently the transformer steps up the voltage for transmission. The loss is dominated by heat losses.	5–20%

Source: International Finance Corporation

Table 4.13 Other PV power plant losses

Loss	**Description**	**Approximate loss factor**
Shading (f_{shade})	Site specific, or far-field shading losses are accounted for in the solar access and shading analysis and do not contribute to PR. Whereas shading that occurs as a result of the actual design of the system (near-field), such as inter-row shading, has a loss factor that contributes to the PR.	<1%
Reflection (f_{ref})	Reflection losses are minimised when the incident light is perpendicular to the module (at STC), however in the field, higher incident angles occur resulting in higher losses.	3–8%
Downtime (f_{down})	Downtime is a period when the plant does not generate due to failure. The downtime periods will depend on the quality of the plant components, design, environmental conditions, diagnostic response time and the repair response time.	Highly variable
Grid availability and disruption (f_{grid})	The ability of a PV power plant to export power is dependent on the availability of the distribution or transmission network. Unless detailed information is available, this loss is typically based on an assumption that the local grid will not be operational for a given number of hours/days in any one year, and that it will occur during periods of average production.	Highly variable
Degradation (f_{degr})	The performance of a PV module decreases with time. If no independent testing has been conducted on the modules being used, then a generic degradation rate depending on the module technology may be assumed. Alternatively, a maximum degradation rate that conforms to the module performance warranty may be considered.	0.5–1.5%/year

Source: International Finance Corporation

The efficiency/derating factors (f) of each loss can be calculated using:

$$f_{\text{LOSS}} = 1 - \text{loss}$$

The PR can then be calculated by multiplying all of the derating factors together.

$$\text{PR} = \sum f_{\text{LOSS}}$$

For example, if all the losses in Table 4.12 are considered, the PR of plant would be calculated:

$$PR = f_{temp} \times f_{dirt} \times f_{man} \times f_{v_{drop}} \times f_{inv} \times f_{trans}$$
$$PR = 0.9 \times 0.98 \times 0.98 \times 0.98 \times 0.96 \times 0.87$$
$$PR = 0.71$$

Annual yield estimation

It is very important to calculate the annual yield of the solar farm as it represents the actual usable amount of energy generated by the system for exporting to the grid. It therefore provides the basis for calculating the annual revenue produced by the system. The energy yield can be estimated using the module's rated power, the available irradiation and the performance ratio:

Once the system yield is estimated, it can be used to determine:

- The amount of energy and hence revenue the solar farm of a given array size will generate; or
- The size of array needed to produce the required amount of energy or revenue from the solar farm.

System performance: supply possibilities

It is important to express the annual yield estimation within a given confidence interval; for example, a P90 value is the annual energy yield prediction that will be exceeded with 90% probability, i.e. there is 10% probability that the P90 value will not be reached. The P50 and P90 energy yield predictions are calculated and provided to investors to represent the assurances provided by the proposed system investment.

While the P90 interval is useful for investors or banks to understand the risks of low yield, it is also useful to determine the risk of undersupply. The output of the solar farm may be limited to a maximum due to network capacity limits or congestion. Lower P-values such as P5 or P10 are useful for quantifying the risk of oversupply circumstances. The array should be sized to ensure that these lower P-values are well above the maximum allowable output of the solar farm.

Annual degradation

Annual degradation refers to the reduction in performance of a PV module year by year as a result of the materials in the module degrading under exposure to light and other environmental conditions. The degradation rate is usually higher in the first year, and then stabilises: a rate of 0.3–1% per annum is typical. Factors that affect the degree of module degradation include:

- quality of materials used in manufacture;

- the manufacturing process;
- quality of assembly and packaging of the cells into the module; and
- level of maintenance at the site.

It is therefore extremely important to choose reputable manufacturers and to carefully review power warranties when selecting system components to meet the system design parameters.

While degradation within the first 6–12 months is caused by initial exposure to sunlight, ongoing degradation may be caused by:

- effect of the environment on the surface of the module (for example pollution);
- discolouration or haze of the encapsulant or glass;
- lamination defects;
- mechanical stress and humidity on the contacts;
- cell contact breakdown; or
- wiring degradation.

It is therefore important to account for annual degradation when predicting PR, system yield and revenue throughout the lifetime of the solar farm.

Variation and sensitivity analysis

When estimating the yield of the solar farm, it is important to consider the inherent intermittency of the solar resource in terms of irradiation, as well as the uncertainty in the methods used to measure it. There are many sources of irradiation variances including spatial uncertainty, year-to-year variability, irradiation data (equipment quality), correlation uncertainty and future variability.

The yield will vary from the long-term average on any given year due to climatic fluctuations making it difficult to predict the energy yield for any given year. The limits of inter-annual variation should be quantified in order to provide a better understanding of the risks to the finance lenders and to perform a sensitivity analysis. This can be done by assessing the long-term radiation data from nearby MET stations or satellites as described above. At least ten years' worth of data should be used in order to give a reasonably accurate assessment of the variation.

Quality control

Quality control should be implemented throughout all stages of planning, design and installation of the solar farm in order to:

- minimise the risk of damages and outages;
- optimise the use of warranties;
- avoid wasting resources;
- optimise the overall performance of the solar farm; and
- ensure compatibility with the proposed network interconnection.

Figure 4.34 This image shows SMA inverter dust testing with particle composition similar to the desert in Arizona. The test showed that no sand or dust got inside the inverter or ventilators, proving it desert proof.

Source: SMA Solar Technology

Product testing

When selecting system equipment (PV modules, inverters, transformers, etc.) during the design stage it can be very beneficial to choose manufacturers that have not only adhered to international testing standards but have implemented tougher requirements such as specific tests for operation in certain climates. For example, SMA conduct tests in a walk-in climate chamber to ensure maximum reliability in extreme climate conditions including:

- temperatures ranging from –40°C to +90°C;
- relative humidity from 10–90%; and
- endurance tests up to 1,000 hours and dust tests with wind speeds up to 20m/s (see Figure 4.34).

Assure product quality and safety

When choosing various components for the design of a solar farm, it is important to ensure the products are safe and reliable. This will involve the choice of manufacturer, but also consideration of site-specific conditions. Some tips for choosing safe, high-quality products are given below. It is also very important to employ best practice guidelines when installing and operating the solar farm in order to maintain the quality and safety of all equipment on-site.

- Choose credible manufacturers based on experience, reputation, resources, qualification, performance guarantee, and product certification.
- Choose products with extensive warranty periods, but first ensure commercial and technical credibility of the manufacturer: the warranty is only valid for as long as the company operates.
- Choose products that have already been used in utility-scale solar systems and review previous experience.

- Ensure the products are suitable for the specific site conditions based on previous testing and experience.
- Ensure the products are certified and comply with local and international quality and safety testing standards.
- Review warranty information including whether or not local technicians are available to service the site if there is a problem or if there is an additional charge for repair or replacement.
- Choose products with detailed specifications or installation/operation manuals to ensure best practices.

Product specific quality benchmarks are described in Table 4.14.

Table 4.14 Quality benchmarks for different components in a solar farm

Product	Quality benchmark
PV module	Manufacturers should provide: • Product guarantee specifying the duration for which the modules will be fully functional (usually between 10–12 years). • Power guarantee specifying the percentage of the module's nominal power that will be delivered throughout different stages of its lifetime (e.g. 96% after the first year, and 80% by the 25th year). • Expected lifetime (typically 25–30 years). • Module should carry IEC or UL certification, testing facility and safety marks relevant to the country of installation.
Inverters	Manufacturers should provide: • Product guarantee specifying the duration for which the inverter will be fully functional (usually around two years). • Inverter mean time between failures figures (usually 5–10 years). • Inverter track record. Inverter protection should include: • Incorrect polarity protection for the DC cable. • Over-voltage and overload protection. • Islanding detection for grid connected systems (depends on grid code requirements). • Insulation monitoring.
Support structures	Manufacturers should provide: • A limited product warranty specifying the duration for which the support structures are fully functional (5–10 years). • The expected useful life (should be at least 25 years). • Steel driven piles should be hot-dip galvanised to reduce corrosion. • Structures should have sufficient load capacity and wind resistance.
Module cables	• Must have wide temperature range (–55 to 125 degrees Celsius). • Must be resistant to UV radiation and weather when laid outdoors. • Must be single core and double insulated. • Must have mechanical resistance (animal proof, compression, tension and bending).
Monitoring systems	Full system monitoring. String monitoring. Data record. System failure notification. Online portal.

Source: International Finance Corporation

Bibliography

Belfiore, F. *Risks and Opportunities in the Operation of Large Solar Plants*. San Diego: Solar Power-Gen, 2013.

Bozicevich, R. 'Solar Module Testing Practices: Future Standards, Current Limitations'. *Solar Industry*, 2013, 2nd edn.

CAT Projects & ARENA. 'Investigating the Impact of Solar'. 2015.

Christiansen, W. and D. Johnsen. 'Analysis of Requirements in Selected Grid Codes'. 2006.

FRV. 'Lessons Learned in the Development of Moree Solar Farm'. 2014.

GSES. 'Chapter 12 – Matching Array and Inverter'. In *Grid-Connected PV Systems: Design and Installation*. 2015.

GSES. 'Chapter 15 – System Efficiency and Yield'. In *Grid-Connected PV Systems: Design and Installation*, 307–31. Sydney, 2015.

GSES. 'Chapter 20 – Commercial and Utility Scale PV Systems'. In *Grid-Connected PV Systems: Design and Installation*, 425–56. Sydney, 2015.

International Finance Corporation. 'Utility Scale Solar Power Plants'. 2012.

Miller, L. and J. Bishop. *Keeping Solar Panels Secure on Unstable Ground.* Oldcastle Precast, 2009.

Solar Assist. *Best Panels.* n.d. www.solarassist.com.au/BestPanelsAU.html (accessed 15 September 2015).

Solar Market. *Choosing the Right Solar System Components.* n.d. www.solarmarket.com.au/tips/choosing_components (accessed 15 September 2015).

Taggart, D. 'Power Factor Control for Grid-Tied Photovoltaic Solar Farms'. Budapest, 2012.

Testa, A., S. De Caro and T. Scimone. 'Sizing of Step-up Transformers for PV Plants through a Probabilistic Approach'. 2013.

Western Power. 'Generator Grid Connection Guide V2: An Introduction to Power Systems and the Connection Process'. 2011.

Wilson, Chris. 'Utility-Scale Solar PV – Grid Integration Challenges and Opportunities'. First Solar, 2015.

Yamada, Hiroyuki. *National Survey Report of PV Power Applications in Japan.* IEA PVPS, 2013.

5

Installation and commissioning

The installation of the solar farm can commence once the system design has been finalised and approved by the necessary parties; the system's testing and commissioning will follow practical completion of the installation process. A high-quality system design is the foundation for a high-quality installation: this design process should incorporate the equipment to be used, the location of the equipment and the configuration of the system.

The installation of a utility-scale PV system requires appropriate project management to deliver the completion of this multidisciplinary work and to ensure that the system is installed according to the applicable standards, best practice guidelines and contracted deliverables.

Installation preparation

In preparation for the installation, it is vital to establish a comprehensive plan with logical sequencing of tasks to ensure the project is constructed on time and within budget. An example of a logical installation sequence is given in Figure 5.1. Each task should be broken up into subtasks and assessed based on the inputs required. More information on each task will be given throughout this chapter.

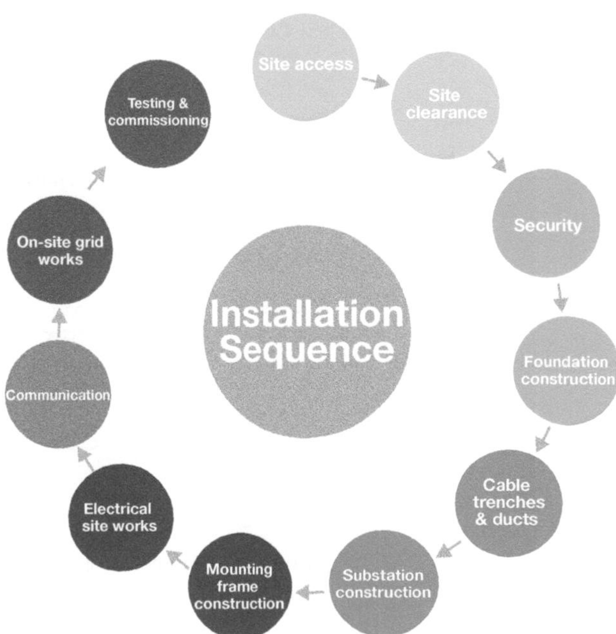

Figure 5.1 A logical sequence of installation works.

Source: Based on information provided by the Institute of Sustainable Futures

Site contractors

There are many different contractors and suppliers from different technical disciplines who are involved in the installation processes, so it is critical to effectively manage the interface between each discipline for the various elements of the installation. It is likely that the EPC (engineering, procurement and construction) contractor or a single lead contractor will remain involved during all the various stages of the installation and they will need to work with personnel from other disciplines. An example of how these project segments may interact is illustrated in Figure 5.2. In preparation for installation, it is important to consider how the interface of these segments will be managed. The interface required will vary depending on the contracting structure and specific requirements of the project. Note that there may be many more project elements and organisations than are shown in Figure 5.2.

Construction plan

In order to ensure that the installation process is completed as efficiently as possible, a detailed construction and management plan should be developed prior to the installation work commencing.

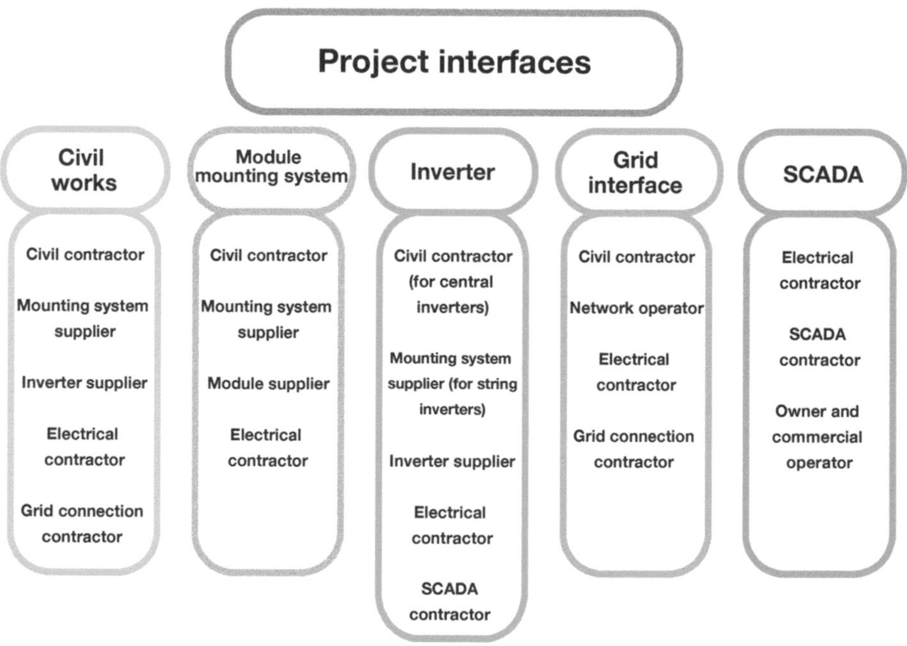

Figure 5.2 Solar farm installation project interfaces.

Source: Based on information provided by the Institute of Sustainable Futures

This plan should include:

- tasks, subtasks and duration of each;
- health and safety (HS) requirements;
- restrictions placed on any task;
- contingency of each task;
- milestones and key dates;
- interdependencies between tasks (helping to define ordering of tasks);
- parties responsible for tasks;
- project critical path; and
- actual progress compared to the plan.

It is imperative that the project's plan incorporates milestones, which are project goals based on contractual obligations, incentives or penalties, as well as fixed dates such as the grid-connection date. If the construction is planned around these milestones, any delays can then be more easily identified and, in response, appropriate decisions about allocating budget and resources can be made.

Cost management

Once construction commences, the project is classified as financially in debt because of the interest and finance costs, but it is not generating any income. It is therefore crucial for the construction of the solar farm to be completed in a timely manner that is consistent with the planned installation programme and that all associated costs are managed carefully.

A cost management plan should include the payment schedule proposed for contractors and manufacturers at various stages of construction. Some payments may be made in advance. Payments to manufacturers are usually paid once the components have been delivered to the site. Payments to contractors may be made on completion of agreed milestones within a specified timeframe, on mechanical and electrical completion of the installation of components or once the system and components have been pre-commissioned and formally commissioned according to the system's technical specifications.

Effective cost management should include a method:

- to determine if the project is running on time and on budget;
- to ensure that work has been completed to a high standard of compliance by all contractors;
- to monitor defects and completion of tests; and
- a thorough budgetary contingency plan.

This may be achieved through:

- Earned-value management
  - Monitors how much time and how much of the budget has been spent on the work completed, relative to the project time and budget plan.
  - Allows project overspends to be estimated and controlled as early as possible.

- Completion certificates
  - Issued once a segment or stage of the solar plant is physically complete and has been installed and commissioned correctly.
  - Payment to the contractor will be withheld until the completion certificate has been issued.
- Defects lists
  - A list of any defects that occur throughout construction.
  - Defects must be addressed and rectified before the project is handed over to the developer.
  - The developer will often accept minor defects if the contractor takes responsibility to rectify them within a specified period.
- System acceptance testing
  - A System Acceptance Test is usually completed by an independent engineer to confirm that industry standard tests have been completed, all non-compliances addressed and the appropriate report provided to the developer.

Site access

In preparation for the construction of the solar farm, new or temporary access roads must be built to accommodate large delivery vehicles as well as pile-driving equipment, excavators, cranes, and lifting equipment. It is estimated that upwards of 20 deliveries by heavy goods vehicles are required per MW of installed capacity of the solar farm. Existing roads may need to be upgraded to accommodate these vehicles travelling to the site. The construction or alteration of access roads should follow the layout in the design plan, aiming to keep access roads to a minimum for aesthetic and ecological reasons. The road plan should be carefully marked out with stakes and flags prior to construction commencing. Depending on the site plan and intended usage, access by road within the solar farm will also be required for security and operations/maintenance over the life of the installation. It is recommended that the proposed access and service roads are planned and built to reflect the intended common usage of these roads over the life of the solar farm.

It is important for the access roads to be constructed with care to minimise erosion and sedimentation in local water bodies. Depending on the extent of access roads required to be built on-site, a range of equipment may be required including:

- angle or fixed blade bulldozer to clear the space for the road;
- a roller to compact-fill material;
- a motor scraper to haul large volumes of displaced soil over large distances; and
- a front-end loader and dump truck to haul earth or rock.

Site clearance

Prior to the construction commencing on-site, earthworks are usually carried out to ensure the site is suitable for installation of foundations and that there is adequate space for equipment delivery and access roads. Earthworks may include:

- land clearance to remove obstructions;
- minor grading and trimming;
- stripping, hauling and stockpiling topsoil and subsoil;
- soil excavation for cable trenching (see Figure 5.3);
- construction of internal site access tracks; or
- installation of on-site erosion and sediment controls, which may include minimising length of time that land is at risk of erosion; installing sediment control devices in parallel with contours of the land; reusing material collected from erosion control and sediment collection on-site; ensuring vehicles remain on existing and access roads when possible; ensuring no earthworks take place just before or after heavy rainfall.

When topsoil and subsoil are stripped for access roads and excavation etc., this material should be stored on-site so that it can be replaced once construction works have been completed.

Earthworks should be kept to a minimum, ensuring that construction impacts are appropriately managed and that techniques have minimal environmental impacts. Preparation of a project-specific environmental management plan for the construction process will ensure that any project hazards are identified and appropriate mitigation measures are adopted during the construction phase. The plan should be in accordance with the requirements of relevant local environmental legislation and guidelines, and should provide adequate information and instruction to on-site workers.

Figure 5.3 Cable trenching Boxted Airfield Solar Farm (UK).

Source: Andrew Stacey

Security

Utility-scale solar plants comprise valuable and relatively portable equipment and are usually remotely located. Therefore, it is important to implement security systems to reduce the risk of theft and tampering of this equipment. It is also critical to ensure that the level of security provided at the site supports the requirements of the insurance policy. Security measures may include:

- Security fences that have been tested and approved by local security standards. This may include fences with barbed wire to prevent people from climbing over it. These fences may also be required for public safety reasons. It is important to reduce the visual impact of the fence by planting shrubs nearby, and to ensure that small animals are able to pass under the fence.
- Electronic security such as CCTV cameras' lights and microwave sensors may be used to trigger an alarm to security personnel in the event of trespass.
- Anti-theft module mounting bolts may be used, where resin is applied once tightened and can only be removed once heated to 300°C
- Anti-theft module fibre may be installed between modules. If a module is removed the fibre breaks and a security alarm is triggered.
- Permanent guarding station with a security guard may be required if area is particularly prone to theft or vandalism.

Figure 5.4 Security fence with barbed wire at Roxboro Solar Farm (North Carolina, USA).

Source: Dickerson Fence

Equipment installation

Foundation construction

When preparing to install the foundations for the PV module array mounting structures, it is important to accurately measure and mark out the location of each pile with a stake, according to the planned layout. This should be done using an accurate levelling device such as a rotating laser (see Figure 5.5). It is also critical to be aware of hazardous obstacles below the surface such as existing cables or pipes, pre-existing foundations or rocks. The main types of module array mounting structures are discussed below.

Earth screws

Earth screws are steel pole shaped structures with threads at the bottom to improve the pull-out resistance of the foundation. If there are no obstruction (ground refusal) issues to this method in the soil, the screws can be installed directly into the ground using a spiral piling excavator, with 360° swing

Figure 5.5 Hilti PR 2-HS rotating laser measuring system, used to level, align, build slopes and square walls or partitions.

Source: Hilti (Aust.) Pty Ltd | hilti.com.au

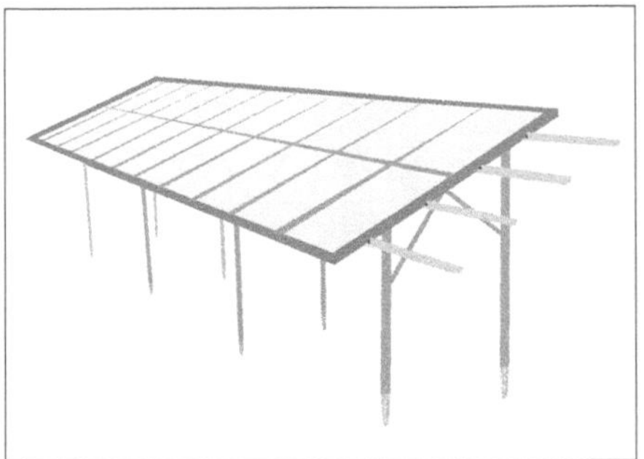

Figure 5.6 Ground screw foundation illustration and construction demonstration.

Source: PowerWay

turntable. Typically 100–200 screws can be installed per day in medium stiff soil. However, if parts of the ground are unable to be penetrated by the screw itself, predrilling will be required to establish these holes with the use of bobcats, excavators or other equipment using auger attachments. This can add significant time and cost to the installation, but is cheaper than attempting to undertake pile-driven installations on sites with high ground refusal. Earth screws may be suitable for sites with soft soils, high wind speeds and rocky terrain.

Helical pile

A helical pile is shaped with a pointed end and a large disc near the end is angled so that when the post is rotated, it wedges into the ground. This type of pile can be easily installed with a hydraulic drill auger attachment on bobcats, excavators or other equipment used to rotate it into the soil. Helical piles have strong pull-out resistance because the disc at the bottom is held down by the weight of soil above it, making it suitable for soft soil types. It is important to determine the embedment depth of the helical pile required to meet the load of PV modules and mounting system. This is done by conducting a pull-out test which measures the vertical and lateral forces at various depths.

Driven piles

Driven pile installations have low pull-out resistance compared to earth screws and helical piles and therefore are more suited to soils with good cohesiveness such as clay, gravel, dense sand and those with low water tables. They can be

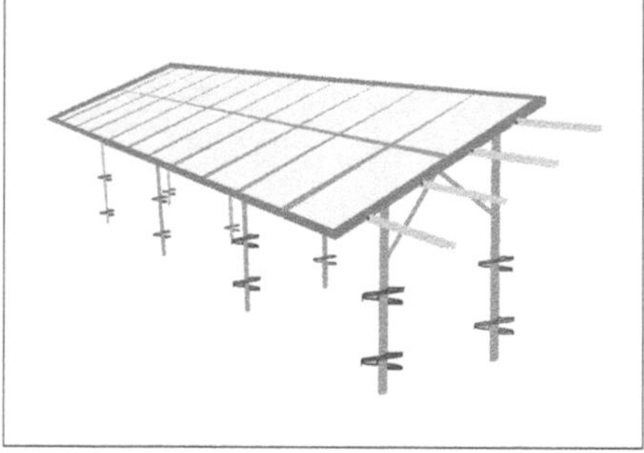

Figure 5.7 Helical pile illustration and construction demonstration.

Source: Larry Reinhart

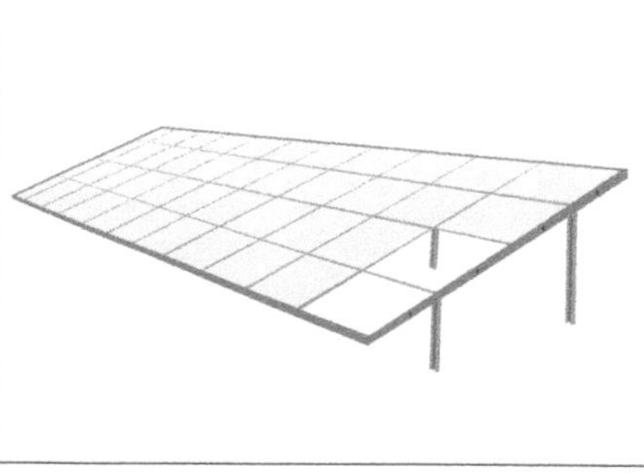

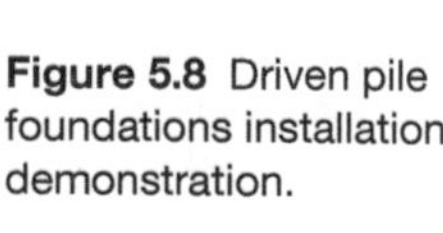

Figure 5.8 Driven pile foundations installation demonstration.

Source: Vermeer

installed quickly and cheaply using pile drivers and can be used on sloped and uneven terrain. If there are ground refusal problems during pile driving, there are three options:

- Conduct a pull-out test to determine whether there is enough resistance at the proposed depth. If proven adequate, cut the pile so that it is the correct length.
- Remove the driven pile and reinstall it nearby assuming that there is enough tolerance in the design of the support structure to allow for this.
- Remove the driven pile, drill an oversized hole, insert the driven pile into the hole, and fill the hole with enough concrete to meet the required pull-out forces.

Ballasts

If the geotechnical survey finds that the soil is difficult to penetrate, is too soft to hold piles, has too many rocks or has subsurface contaminants, then bolting the frames onto ballasted foundations may be the most suitable option. These are also a safe option if a geotechnical survey was not carried out at all. Ballasted foundations are very simple and easy to work with – consisting of precast concrete footings that sit above the soil (Figure 5.9). However, care must be taken to allow for soil settling, erosion and heaving. Additionally, ballasted footings have limited use depending on the topography of the site: they may not be suitable for sloped or uneven terrain. Ballast blocks are lifted on-site using bobcats or excavators and placed into pre-marked locations on-site. It is important to ensure that all ballasts are level. The frame is then bolted onto the ballasted foundation, with a tilt arm set to the angle determined in the design. Alternatively, ballast blocks can be set in a tray to hold the mounting frame down.

Mounting frame construction

Once the foundations have been constructed, the mounting frames can be installed on these foundations. While there are different types of mounting frames, a very common structure is illustrated in Figure 5.10. This is a single row, vertical post design, which reduces the number of ground penetrations compared to double row, and can be installed without the need for any lifting equipment or machinery. Details on the installation procedure are given in Figure 5.11.

Some alternative choices for mounting frame structures are given in Figure 5.12.

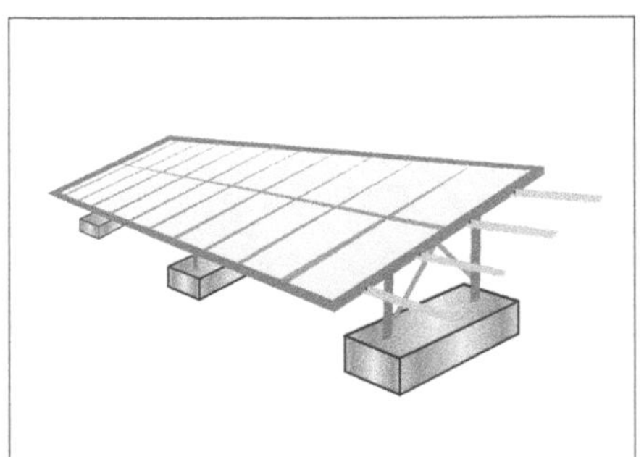

Figure 5.9 The figure on the left illustrates a concrete ballast foundation where the frames are attached directly to the ballast block. The figure on the right illustrates ballast blocks being installed using a forklift.

Source: Jim Young

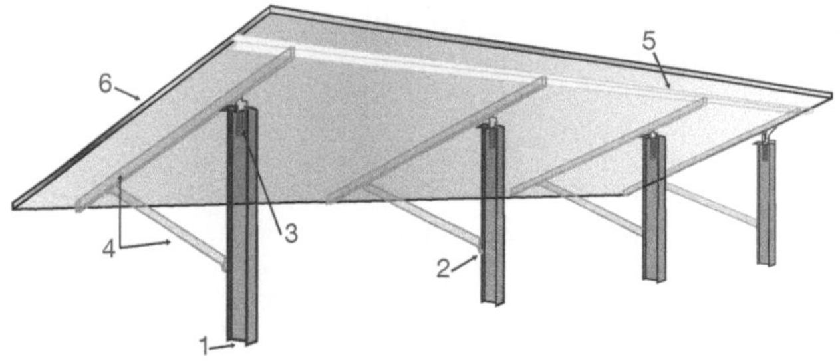

Figure 5.10 Mounting frame structure.

1. Standard I beams
• Pile driven with standard equipment

2. Strut attachment
• Used to attach the strut to the foundation
• Field adjustable
• Should be hot dip galvanised for corrosion protection

3. Strongback attachments
• Used to attach the strongback to the foundation
• Should be hot dip galvanised for corrosion protection

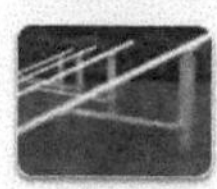

4. Strongback assembly
• Assemble the strongback strut and rail brackets (often factory pre-asembled)
• Unfold and attach to the foundation

5. Module rails
• Installed by hand-lifting into place
• Should be lightweight and high strength
• Should have built in wire channels

6. Module clamps
• Pre-install clamps then slide modules into place
• Ensure end clamps are installed at the end of modules, and that mid-clamps are installed in between modules

Figure 5.11 Mounting frame construction procedure.

Source: Preformed Line Products Company

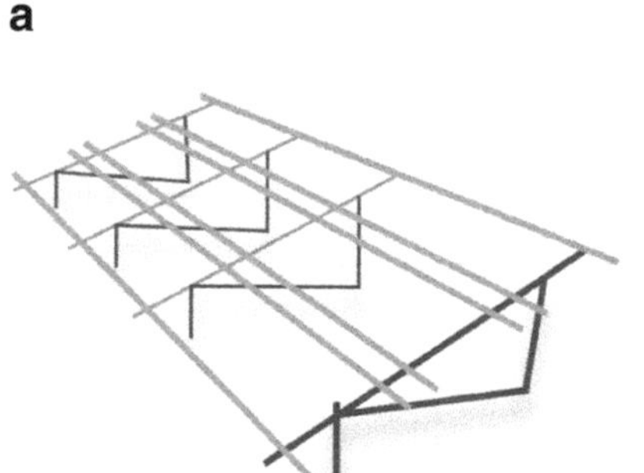

Figure 5.12 (a) N-type mounting structure; (b) typical double row, vertical post design; (c) single-axis tracker; (d) dual-axis tracker mounting system.

Source: Global Sustainable Energy Solutions

Attaching modules

The most common method of attaching the modules to the mounting system is to use clamps, but it is important to install them correctly. Clamps should:

- be the correct size according to the module frame height;
- be manufactured from the correct material; and
- extend only over the module frame – they should not cover any of the PV cells or put pressure on the glass.

In addition:

- Use end clamps to secure the outside edges of the first and last module and mid-clamps for the inside edges between the modules; using incorrect clamping can damage the module and can affect the integrity of the installation.
- Clamps should be installed so that they are square (i.e. parallel) with the module and square with the mounting rail.
- Clamps should be spaced according to the manufacturer installation instructions. If there is too much space between clamps on the edge of the module, there may not be enough support in the middle; if there is too little space between clamps, the ends of the module may be unsupported.

Other methods of attaching modules to mounting structures include using pre-drilled mounting holes (Figure 5.13a) or an insertion system (Figure 5.13b). The insertion system can allow for quicker and easier installation compared to clamps, and can avoid damage to the system by allowing installation to take place from below the modules without putting pressure on from above.

It is important to avoid removing, modifying or drilling into the module frame as this may void the manufacturer's warranty.

a

b

Figure 5.13 (a) Well-installed clamp: correct use of mid-clamp between modules, clamp is square with module, bolts are made of galvanically similar metal. (b) Poorly insulated clamp: incorrect use of end-clamp between modules, clamp is not square with either the module or the rail, the bolt has corroded because it is not galvanically similar to the surrounding metals.

Sources: (a) Jack Parsons; (b) Global Sustainable Energy Solutions

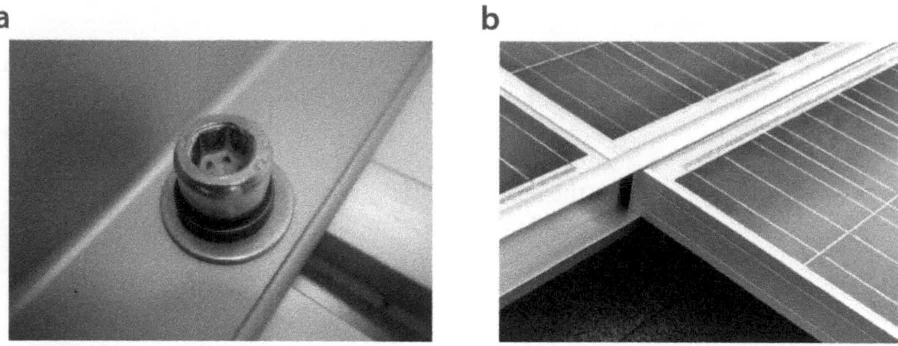

Figure 5.14 (a) Attachment using pre-drilled frame mounting holes. (b) Attachment using insertion system, which can be quicker and easier and safer to install than when using clamps.

Source: LDK Solar

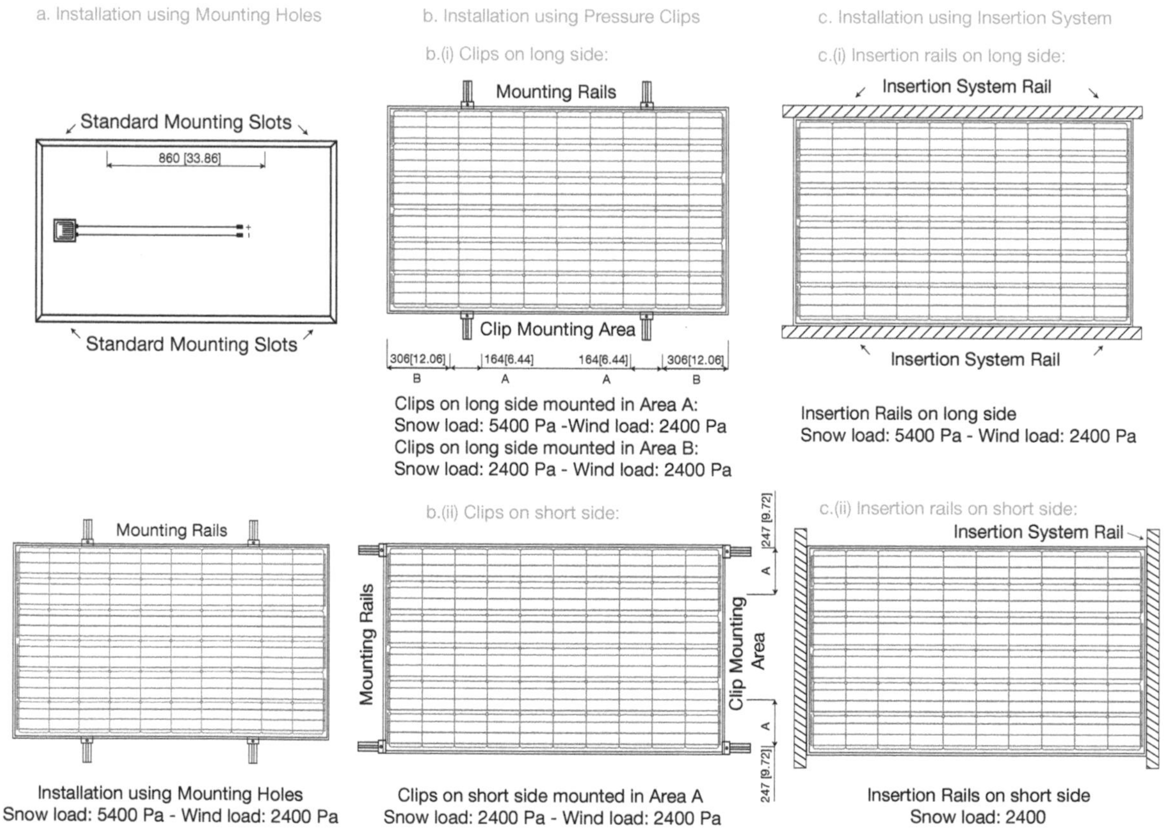

a. Installation using Mounting Holes

Standard Mounting Slots

860 [33.86]

Standard Mounting Slots

Mounting Rails

Installation using Mounting Holes
Snow load: 5400 Pa - Wind load: 2400 Pa

b. Installation using Pressure Clips

b.(i) Clips on long side:

Mounting Rails

Clip Mounting Area

306[12.06] 164[6.44] 164[6.44] 306[12.06]

B A A B

Clips on long side mounted in Area A:
Snow load: 5400 Pa -Wind load: 2400 Pa
Clips on long side mounted in Area B:
Snow load: 2400 Pa - Wind load: 2400 Pa

b.(ii) Clips on short side:

Mounting Rails

Clip Mounting Area

247 [9.72] A A 247 [9.72]

Clips on short side mounted in Area A
Snow load: 2400 Pa - Wind load: 2400 Pa

c. Installation using Insertion System

c.(i) Insertion rails on long side:

Insertion System Rail

Insertion System Rail

Insertion Rails on long side
Snow load: 5400 Pa - Wind load: 2400 Pa

c.(ii) Insertion rails on short side:

Insertion System Rail

Insertion Rails on short side
Snow load: 2400

Figure 5.15 Comparison of three different mounting methods, including (a) mounting holes, (b) clamps and (c) an insertion system.

Source: LDK Solar

Inverter installation and transportation

When installing the inverters, it is very important to follow the guidelines given in the manufacturer's installation manual. The physical dimensions of central inverters used in utility-scale systems are much larger than those used in residential or commercial size systems. They require correct equipment transport, lifting and installation. An inverter can weight up to 4 tonnes and more than 30 tonnes if it includes a MV-transformer and MV switchgear. The inverter may be transported to the site using a forklift, pallet truck with a crane fork to lift it on to the truck for transport to the site; this same equipment may also be used to lift it from the truck into its final position, but often a crane is used (see Figure 5.16). It is important that all lifting equipment is designed to take the weight of the inverter. Care must be taken during all stages of transportation, following the guidelines set out in the inverter's installation manual and the established HS requirements.

It is important that the inverter is set on suitable foundations, such as foundation slabs or platforms, strip foundations, or pile-driven steel pillars set on dry solid foundations such as gravel. The foundations and platforms must be suitable for the weight of the inverter.

There should be adequate access areas and routes to enable easy installation, maintenance and replacement of the inverter. According to SMA's installation manual, the clearance areas for servicing should be:

- permanently vehicle accessible;
- constructed of well-compacted earth, gravel, tar, concrete or similar material;

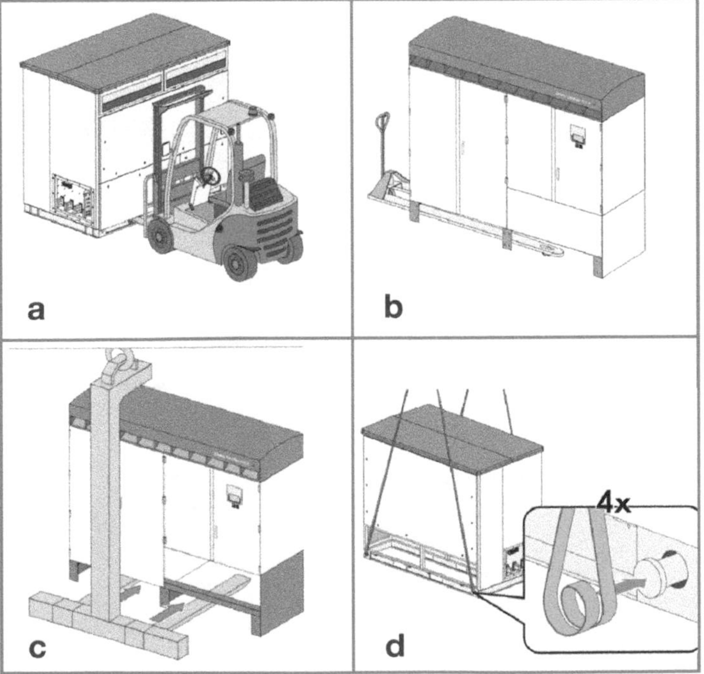

Figure 5.16 Inverter lifting methods: (a) forklift, (b) pallet truck, (c) crane fork and (d) crane.

Source: SMA Solar Technology

- suitable for use with lifting aids, truck mounted cranes, and forklift trucks;
- free of obstacles such as fences and barriers;
- of a maximum grade of 3%;
- even with the foundation; and
- compliant with local, state and federal occupational health and safety legislation.

Figure 5.17 and Figure 5.18 illustrate foundation options and minimum clearances according to SMA's installation manual.

Electrical works

Module wiring

The modules should be positioned according to the system design. Normally modules that are in the same string should have the same orientation and tilt; however, where micro-inverters or power conditioning units on each module are used, this may mean that modules do not need the same orientation and/or tilt.

Modules are usually connected together using the pre-installed cables and multi-contact (MC) connectors on each module. These cables are wired inside the module junction box located on the back of the module as shown in Figure 5.19. The array wiring, as with all wiring, should be installed so that it is neat and tidy, and mechanically protected from damage using UV-rated conduit (see Figure 5.21a) or cable trays, so that relevant wiring standards are met. All cables and connectors used must be appropriately rated, as outlined in Chapter 4.

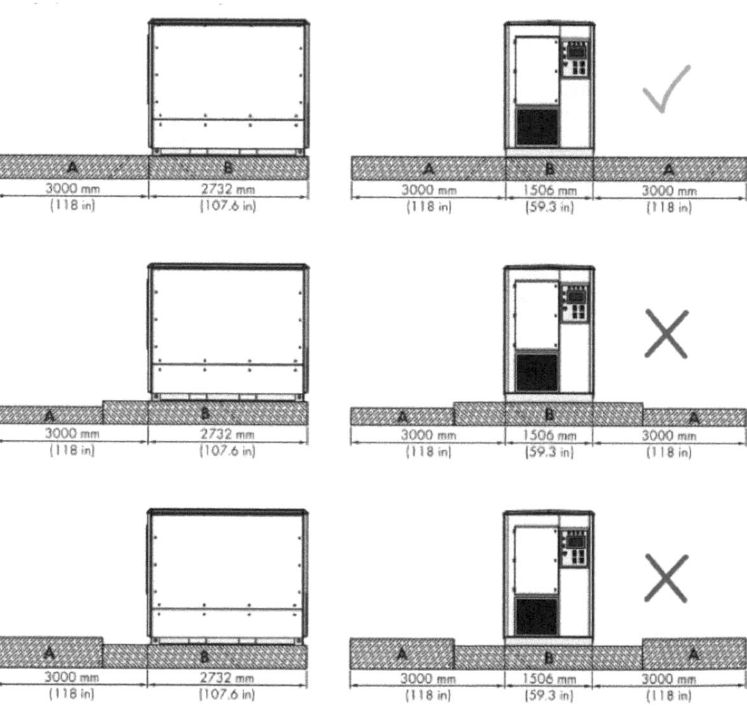

Figure 5.17 Mounting options: foundations and minimum clearances, where A represents clear spaces and B represents foundations or platform.

Source: Installation Manual: Sunny Central 1850-US, Sunny Central 2200-US, SMA Solar Technology

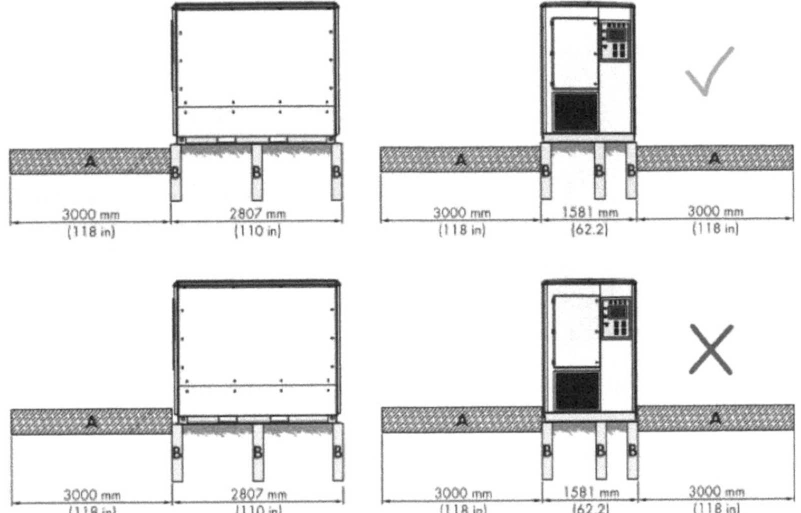

Figure 5.18 Mounting options: pile-driven steel pillars and minimum clearances, where A represents clear spaces and B represents pile-driven steel pillars.

Source: Installation Manual: Sunny Central 1850-US, Sunny Central 2200-US, SMA Solar Technology

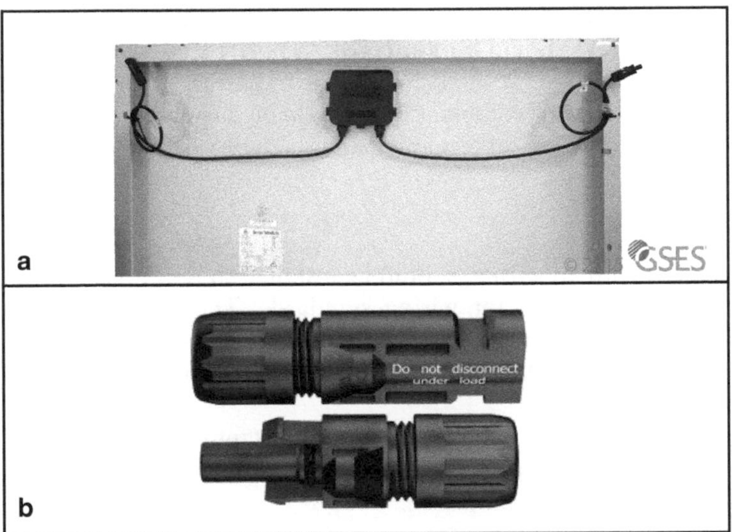

Figure 5.19 (a) Pre-wired array cables at the back of the module with plug-and-socket connectors; (b) Multi-contact MC4 male (above) and female (below) connectors used to extend and connect module cables. It is also important for the modules to be wired in a way that avoids conductive loops (Figure 5.20). Reducing conductive loops will lower the risk of lightning-induced over-voltages in the array wiring, as well as reducing interference to AM/FM radio signals.

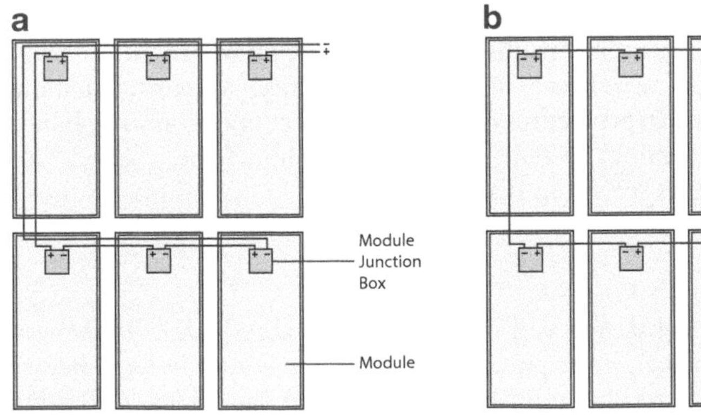

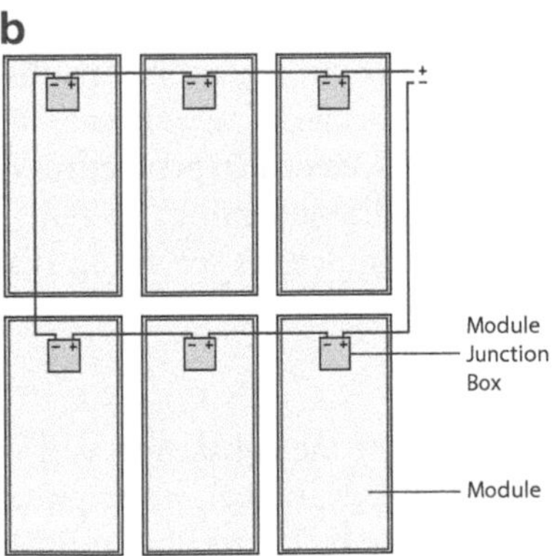

Figure 5.20 (a) PV string cable with minimised inductive loop area. (b) PV string cable that has an inductive loop. This should be avoided.

Source: Global Sustainable Energy Solutions

a

b

Figure 5.21 (a) Poorly installed cables and (b) well-installed cables kept tidy and protected with conduit.

Source: Solarpraxis Neue Energiewelt AG

For safety reasons, the connection of the modules together in a string should be performed after the DC switch-disconnector has been installed and connected.

The strings and sub-arrays should be combined using a string combiner box to house the system protection. It must have appropriate IP rating and UV resistance. All cable entries into a combiner box should be through the allocated base entry points to prevent water ingress and maintain the enclosure's IP rating.

Inverters

The inverter wiring, as with all wiring, should be installed so that it is neat and tidy and complies with the relevant wiring standards. The cables must be appropriately rated, as outlined in Chapter 4, and capable of withstanding high temperatures up to 90°C. If conduit is used, it should be rain-tight, waterproof and UV resistant.

When connecting the cables to the inverter, it is important to maintain the IP rating of the inverter by following the manufacturer's installation instructions carefully.

Transformer substation

The construction of the transformer substation includes civil works, installation of cables, transformers, switchgear, lightning protection systems, earthing protection systems, electrical protection systems, and grid connection compliance protection and disconnection systems. Installation should be carried out according to the guidelines set out in IEC/TS 61936.

Civil works

The construction of the substation will involve site clearance (detailed below in the section on site clearance), in preparation for setting the foundations. Before the foundations are built, trenches must be constructed for the cables connecting the inverters to the substation. A pit is dug out for the cables entering directly below the substation. Conduits are then laid in the trenches and up through the pit. Steel reinforcement for the concrete slab foundations should be installed. Cement is then poured into the pit and around the conduit and should be allowed sufficient time to cure (usually 28 days). The conduits should be cleared with compressed air so the cables can be pulled through. This is done by first pushing a rope through the conduits and tying it to the cables, an electric winch is then used to pull the cables all the way through from the inverters to the substation (Figure 5.23). Once the cement has cured and the cables have been pulled through the conduits, the transformer and switchgear can be installed on to the foundations using a crane, ensuring an upright position is maintained (Figure 5.22).

Transformers

When connecting the transformer terminals to incoming and outgoing conductors, it is very important to follow the diagrams and instructions provided by the manufacturer. Some recommendations include:

- torque requirements for the nuts and bolts as specified by the manufacturer;
- using copper or aluminium cables;
- properly sized, mechanical or compression-type lugs with appropriate safety marks should be used as specified by the manufacturer;

Figure 5.22 Installation of two 15 kV metal-enclosed switchgear assemblies at Five Points Solar Farm in California, USA.

Source: Courtesy of Blue Oak Energy, LLC

Figure 5.23 Workers at Five Points Solar Farm in California, USA, use an electric winch to pull MV cables into one of the site's metal-enclosed switchgear assemblies.

Source: Courtesy of Blue Oak Energy, LLC

- lugs should be attached to the cables as specified by the termination or cable manufacturer;
- washers should not be installed between terminal lugs and the termination bus bar; and
- cables should be rated to either the nameplate capacity of the transformer of 125% of the nameplate capacity, however oversizing is more expensive and often unnecessary.

Switchgear

The installation, cabling and protection of the switchgear depend greatly on the size and type of switchgear chosen. The instructions provided by the manufacturer and the local standards should be carefully followed.

Cables

MV (medium-voltage) cable terminations serve a number of important functions apart from enabling electrical connection, including:

- Relieving voltage stress that would otherwise build up at the insulation shield termination point.
- Sealing the termination against moisture and environmental contaminates.
- Preventing electrical treeing or tracking at the termination point.

It is therefore very important to terminate MV cables properly, which requires specialised skills and tools. Since MV cable termination methods and procedures vary by manufacturer, it is important that the installers receive training specific to the product and follow the manufacturer's instructions.

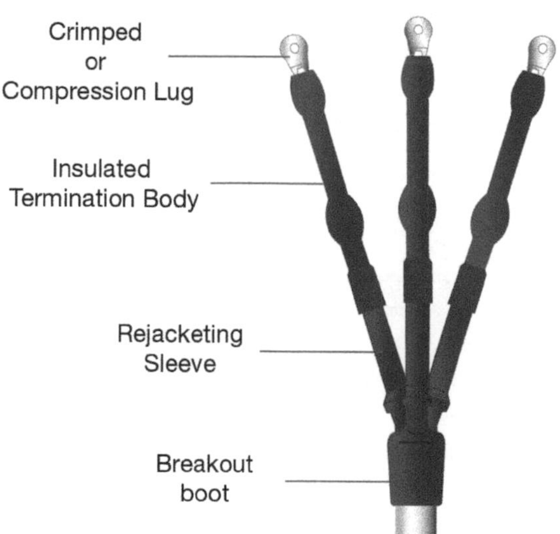

Figure 5.24 Typical MV cable termination.

Earthing protection and short-circuiting equipment

According to IEC/TS 61936, each component of the substation that can be isolated from the system should be provided with a means of earthing and short-circuiting. Earthing and short-circuiting devices provide a means of short circuiting the live conductors to earth, during periods when a particular component must be isolated from the system. An earth grid is typically installed adjacent to the substation installation. Earthing and short-circuiting devices may include:

- earthing switches (preferably fault-making and/or interlocked);
- earthing switch trucks;
- earthing equipment integrated with other switching devices e.g. circuit-breakers;
- free earthing rods and short-circuiting equipment; and
- guided earthing rods and short-circuiting equipment.

Electrical protection systems

According to IEC/TS 61936, the installation of the substation should implement a number of measures to protect personnel against dangers resulting from arc fault. These include:

- Protection against operating error
  - load break switches instead of disconnectors
  - short-circuit rated fault-making switches
  - interlocks
  - non-interchangeable key locks
- Operating aisles as short, high and wide as possible
- Solid covers as an enclosure or protective barrier instead of perforated covers or wire mesh
- Equipment tested to withstand internal arc fault instead of open-type equipment
- Arc products to be directed away from operating personnel, and vented outside the building, if necessary
- Use of current-limiting devices
- Very short tripping time; achievable by instantaneous relays or by devices sensitive to pressure, light or heat
- Operation of the plant from a safe distance
- Prevention of re-energization by use of non-resettable devices which detect internal equipment faults, incorporate pressure relief and provide an external indication.

Monitoring equipment

The SCADA remote telemetry unit (RTU) racks must be installed according to the design drawing. RTU configuration files will then need to be loaded into the RTU. Similarly, the local control facility (LCF) must be installed either on the desktop or in the allocated panel, as indicated on the drawings. The LCF will then need to be programmed according to the local grid requirements.

Signage and labelling

Clear and unambiguous signage and labelling are required under IEC/TS 61936 to avoid incorrect operation, human error and accidents during operation and maintenance. All nameplates, signs and labels should be made of durable and non-corrosive material and printed with indelible characters.

The following signage is required:

- Information and warning plates on the outside of the room and on each access door to all electrical equipment and operating areas.
- Electrical hazard warning signs on all sides of outer perimeter fences and masts, poles and towers with a transformer or switching device.
- Warning label on capacitors indicating discharge time.
- Emergency signs for emergency exits.
- Cable identification marks inside the transformer cable boxes, immediately outside the transformer cable boxes (single-core cables), inside the indoor switchgear or the outdoor ring main units, immediately outside the indoor switchgear or outdoor ring main units (single-core cables).

Testing and commissioning

Once the solar farm has been installed, it must pass a series of commissioning tests to ensure it is structurally and electrically safe and robust, that it operates as designed and performs as expected. It should comply with IEC/TS 62446 or other relevant local standards and guidelines. Commissioning tests are normally split into three stages: visual inspection, pre-connection system testing and post-connection system testing.

Inspection

A visual inspection of all significant aspects of the plant should take place before the installation is energised. The following guidelines are based on IEC/TS 62446, to check that the system has been installed according to IEC/TS 60364 in general and IEC/TS 60364-7-712 in particular.

DC system inspection

- the DC system has been designed, specified and installed according to the standard;
- all DC components are rated for continuous operation at DC and at the maximum; possible DC system voltage and maximum possible DC fault current;
- protection by use of class II or equivalent insulation adopted on the DC side;
- PV string cables, PV array cables and PV DC main cables have been selected and installed to minimise the risk of earth faults and short circuits; typically

achieved by the use of cables with protective and reinforced insulation (double insulation);

- wiring systems have been selected and installed to withstand the expected site conditions such as wind, ice formation, temperature and solar radiation;
- for systems without string over-current protective device: verify that the module's reverse current rating is greater than the possible reverse current; also, verify that the string cables are sized to accommodate the maximum combined fault current from parallel strings;
- for systems with string over-current protective device: verify that the string over-current protective devices are fitted and correctly specified to local codes or to the manufacturer's instructions for protection of PV modules;
- verify that a DC switch-disconnector is fitted to the DC side of the inverter;
- if blocking diodes are fitted, verify that their reverse voltage rating is at least $2 \times V_{oc@STC}$ of the PV string in which they are installed;
- if one of the DC conductors is connected to earth, verify that there is at least simple separation between the AC and DC sides and that earth connections have been constructed to avoid corrosion.

Protection against overvoltage/electric shock

- Verification of type B residual current device (RCD)/ground fault detection interrupter (GFDI) where installed: an RCD/GFDI is required when the PV inverter is without at least simple separation between the AC side and the DC side; codes will provide further information on when RCDs/GFDIs are required;
- To minimise voltages induced by lightning, verify that the area of all wiring loops has been kept as small as possible;
- Where required by local codes, verify that array frame and/or module frame protective earthing conductors have been correctly installed and are connected to earth; where protective earthing and/or equipotential bonding conductors are installed, verify that they are parallel to, and bundled with, the DC cables.

AC system

- A means of isolating the inverters have been provided on the AC side;
- All isolation and switching devices have been connected such that PV installation is wired to the 'load' side and the public supply to the 'source' side;
- The inverters' and transformers' operational parameters have been programmed to local regulations.

Substation

Inspection of the substation should be carried out to verify compliance of the installation with IEC/TS 61936 or relevant local standards and compliance of the equipment with the applicable technical specifications. This includes:

- verification of characteristics of the equipment (including rated values) for the given operating conditions;
- verification of minimum clearances between live parts and between live parts and earth;
- verification of minimum heights and of protective barrier clearances;
- visual inspections of electrical equipment and parts of the installation;
- inspection of markings, safety signs and safety devices;
- verification of correct fire ratings for buildings/enclosures;
- verification that emergency exits are operational; and
- verification of the earthing system.

Labelling and identification

- All circuits, protective devices, switch-disconnectors and terminals are suitably labelled.
- All DC junction boxes (PV generator and PV array boxes) carry a warning label indicating that active parts inside the boxes are fed from a PV array and may still be live after isolation from the PV inverter and public supply.
- The main AC isolating switch-disconnector is clearly labelled.
- Dual supply warning labels are fitted at point of interconnection.
- A single-line wiring diagram is displayed on-site.
- Inverter protection settings and installer details are displayed on-site.
- Emergency shutdown procedures are displayed on-site.
- All signs and labels are suitably affixed and durable.

Pre-connection system testing

Testing of the electrical installation must be carried out to ensure that the system operates and performs as intended. Testing should adhere to the requirements of IEC/TS 60364-6; measuring instruments and monitoring equipment and methods should be chosen in accordance with the relevant parts of IEC/TS 61557.

The following tests should be carried out prior to grid connection in the following order:

- Tests to all AC circuits to the requirements of IEC/TS 60364-6.

Once tests to the AC circuits are complete, the following tests shall be carried out on the DC side:

- Continuity of protective earthing and/or equipotential bonding conductors, where fitted
- Polarity test
    - once confirmed, all cables should be checked to make sure they are connected correctly to system devices such as switching devices or inverters;
- String open-circuit voltage test
    - should be done before closing any switches or installing string over-current protective devices;
    - measured V_{OC} should be compared with the expected value;

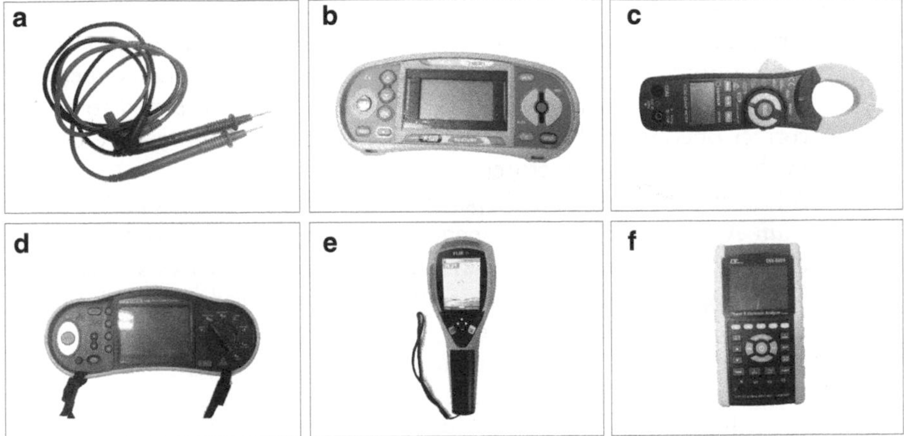

Figure 5.25 (a) Test leads are important for making easy, safe and accurate measurements. Therefore, it is important to ensure that the test leads are in good condition. (b) The MI 3108 InstalTestPV is a multifunction instrument for I–V curve test of PV strings and modules. I–V curve tracers are essential for gathering data to quantify and monitor system performance. (c) 400A AC/DC clamp meter read up to 600 V DC and 400 V AC and 400 A (in both AC and DC). (d) The Fluke 1652 multifunction tester can measure insulation resistance up to 50 GΩ, using 250 V, 500 V or 1,000 V test voltages. (e) Infrared cameras (pictured is the FLIR i5) can identify high-resistance electrical connections and thermally stressed overcurrent protection devices. They also help identify module issues such as cracked cells, faulty internal connections and defective bypass diodes. (f) A 3 Phase Power & Harmonic Analyser – DW6095 is used to do basic power analysis, commissioning and fault finding. It has features like current, voltage, real power, apparent power, reactive power, phase angle and harmonic measurements.

Source: multiple

- String short-circuit current test
  - should be done to verify that there are no major faults within the PV array wiring;
  - measured I_{SC} should be compared with the expected value;
- Substation tests (IEC/TS 61936)
  - voltage test for cables;
  - power frequency voltage test for switchgear;
  - functional tests of electrical equipment and parts of installation;
  - functional tests and/or measuring of protective, monitoring, measuring and controlling devices.

Post-connection system testing

The following tests should be carried out after grid connection in the following order:

- PV string operational test
  - Current should be measured with the system switched on and in normal operation mode;
  - Measured values should be compared with the expected value;

- Functional tests
  - Switchgear and other control apparatus should be tested to ensure correct operation and that they are properly mounted and connected;
  - All inverters forming part of the PV system shall be tested to ensure correct operation; the test procedure should be the procedure defined by the inverter manufacturer;
  - A loss of mains test shall be performed: with the system operating, the main AC isolator shall be opened – it should be observed (e.g. on a display meter) that the PV system immediately ceases to generate; following this, the AC isolator should be re-closed and it should be observed that the system reverts to normal operation;
- Insulation resistance of the DC circuits
  - Follow the insulation resistance test device instructions to ensure the test voltage is in accordance with Table 5.1 and readings in MΩ;
- Performance ratio test
  - Measured PR should be compared with the value stated in the contract;
- Availability test
  - Measures the percentage of time that the plant is generating electricity;
  - Usually carried out over a period of five days.

Verification records

Verification information and results should be recorded and made available to the developer and owner.

The system verification documentation should include the following as a minimum:

- List of equipment supplied
  - All system components and quantities;
  - Information on all equipment including: brand, model/type, serial number;
- System wiring diagrams
  - A basic circuit diagram (single-line diagram) that includes the electrical ratings of the PV array and inverter;
  - A plan view drawing of the installed system showing the location of all the major components;
  - Detailed wiring diagrams;

Table 5.1 Minimum values of insulation resistance according to IEC 62446

System voltage ($V_{oc\ stc}$ × 125) V	Test voltage V	Minimum insulation resistance MΩ
<120	250	0.5
120–500	500	1
>500	1,000	1

- System performance estimate
  - Expected energy yield;
  - Performance ratio;
  - Estimated revenue;
  - Assumptions made to estimate these values;
- Operating instructions – system and components
  - Overview of system;
  - Function of all components;
- Shutdown and isolation procedure for emergency and maintenance
  - How to isolate all or part of the system when maintenance is being performed;
  - How to isolate and shut the system down in an emergency;
- Maintenance procedure and timetable
  - Details on maintenance procedure and timetable given in Chapter 6;
- Monitoring of system
  - Manual on how the SCADA and other monitoring systems work and how to read the outputs;
- Warranty information
  - Product warranties covering defects in manufacture;
  - Product warranties related to product output performance over time;
  - System warranties relating to the installation of the system and its performance over time: usually the installer will document the 'installation warranty' provided, e.g. a minimum of 12 months;
  - Energy performance warranties relating to the guaranteed energy output of the solar farm over a period of time;
- Equipment manufacturers' documentation and handbooks
  - Inverter manuals;
  - Transformer manuals;
  - PV module data sheets;
  - SCADA manual;
  - Product certification;
  - Technical information on balance of system equipment;
  - Mounting system data sheet;
  - Structural certification;
- Testing records and installation checklist
  - Detailed records of all testing should be included;
  - Detailed installation checklist.

Bibliography

Bray, C. 'Ground-mounted PV'. *Solar Pro*, June/July 2010.

BRE National Solar Centre. 'Planning Guidance for the Development of Large Scale Ground Mounted Solar PV Systems'. 2013.

Bushong, S. 'White Paper: Foundation Selection For Ground Mounted PV Solar Systems'. *Solar Power World*, 2014.

DPW Solar. 'Power Peak AL-Large Scale Ground Mount System'.

Ergon Energy. 'Substation Construction Manual'. 2014.

First Solar. 'Construction Environmental Management Plan Nyngan Solar PV Power Station'. 2014.

GSES. 'Chapter 16 – System Installation'. In *Grid-Connected PV Systems: Design and Installation*, 333–59. Sydney, 2015.

GSES. 'Chapter 17 – System Commissioning'. In *Grid-Connected PV Systems: Design and Installation*, 361–82. Sydney, 2015.

International Finance Corporation. 'Utility Scale Solar Power Plants'. 2012.

Lock Solar. *Ground Mount Solar Solutions*. n.d. http://locksolar.com/ground-mounting (accessed 31 August 2015).

Morgan, R. 'Guidelines for Installing Transformers, Part 1'. *Electrical Construction and Maintenance*, 1999.

Pickerel, K. *No Clamps? No Problem*. 4 October 2014. http://solarbuildermag.com/featured/creotecc-clampless-mounting (accessed 22 September 2015).

PV Powerway. *Solar Mounting Powerscrew*. 2011. www.power-groundscrew.com/en/products.html#installation (accessed 28 September 2015).

Simpson, D. 'Basics of Medium-Voltage Wiring'. *Solar Pro*, 2013.

SMA. 'Installation Manual: Sunny Central 1850–US, Sunny Central 2200–US'. n.d.

SMA. 'Installation Requirements: Sunny Central 1850–US, Sunny Central 2200–US'. n.d.

SnapNrack Solar Mounting Solutions. 'Series 200 UL Ground Mount System'. n.d.

URS. 'Baralaba Solar Farm Development'. 2014.

USDA. *The Layman's Guide to Private Access Road Construction*. Haywood Press, 2005.

Wolfe, P. *Solar Photovoltaic Projects in the Mainstream Power Market*. Routledge, 2013.

6

Operation and maintenance

The operation and maintenance (O&M) of a solar farm is intended to deliver both practical and financial outcomes: the system will have prescribed financial criteria to meet over the life of the project and these criteria will be inextricably linked to the performance of the solar farm and the power delivered to the network. As with any piece of equipment, particularly equipment having a specified performance quotient, the scope of the solar farm's O&M will be included as part of the contractual deliverables. O&M will vary according to the structures applied to:

- How and where the system's operation is physically monitored.
- What provisions have been specified by the performance contract regarding how often and how quickly equipment malfunction or failure must be remedied.
- What training is required before system handover for all personnel responsible for system monitoring, maintenance and equipment repairs/replacements.

This chapter outlines the personnel, processes and tasks typically used for the O&M of a solar farm.

O&M planning

O&M activities focus on maximising revenue and mitigating risk to prevent the loss of revenue that may arise from loss of system efficiency, system faults, equipment downtime or other causes.

Often the system owner will utilise the services of an O&M specialist contractor. The agreement between the system owner and the O&M contractor must clearly define the scope of services covered, the associated costs and the responsibilities of both parties. It is imperative that all aspects of O&M are specified as part of the project planning documents from its inception as these criteria are tailored according to the specific operational parameters of the system. Typical O&M planning documents include:

- normal operational procedures;
- safety and security plan;
- environmental protection plan;
- spare parts plan;
- warranty repairs plan;
- maintenance programme plan; and
- performance and availability plan.

Normal operational procedures

Normal operational procedures will define the roles of the personnel required to operate and maintain the system. Depending on the size, location, complexity and network requirements of the system, O&M contractors may choose to place personnel on-site or to manage the plant remotely. The normal operation of the system will also depend on the intended revenue stream for the energy generated. For example, the system may operate differently depending on whether it is competing in an electricity market, is the subject of a power purchase agreement (PPA) having a contracted purchaser, is allocated a feed-in tariff, or if it has some other operationally dependent revenue structure. The normal operational procedures will also define how the system interacts with the local transmission network, what operators are expected to do in order to ensure that the plant meets grid requirements and can provide grid support.

Safety and security plan

The safety and security plan identifies and addresses all site safety and security risks. A site safety and security assessment should be undertaken, and a risk analysis performed. The risk analysis will describe each risk in detail and will quantify the likelihood of its occurrence and the severity of the consequence. The mitigation measures will then be determined and specified for each specific risk. The result of this analysis is the safe work method statements (SWMS) for all O&M aspects of the system. SWMS define what activities are to be carried out in order to ensure the safety and security of both site personnel and operational equipment.

Environmental protection plan

The environmental protection plan will address the handling of site materials and the disposal of waste products, with a particular focus on any toxic or otherwise harmful materials. This plan will also define how to manage local flora and fauna issues and should include any risks identified from the environmental impact statement, which would have been completed during the planning phase, as discussed in Chapter 3.

Spare parts plan

The spare parts plan will identify which major items and which balance-of-system equipment must be kept on-site in the event of product failure. The plan specifies how inventory figures are maintained and at what inventory stock levels are required. Depending on the way the O&M contract is structured, this plan may also include a cost–benefit decision matrix for replacing faulty equipment. For example, it may cost more for a service technician to make a special trip to service a single string fault than the benefit it would add to the system revenue; it may make more sense to wait until the next scheduled maintenance. Conversely, if the system were to have a central inverter fault, this is likely to have a significant impact on system availability and performance; in this case, the required spares would be allocated immediately for the rectification work to take place.

Warranty repairs plan

The warranty repairs plan is related to the spare parts plan in that any major systems equipment incorporated in the system should have a workmanship and performance warranty associated with it. The warranty repairs plan includes the process for carrying out a warranty repair as well as specifying which information must be provided to the equipment manufacturer in order to submit a warranty claim. This plan will also specify the roles of the system owner and the O&M contractor regarding the processing of warranty claims and collection of damages and/or replacement parts.

Maintenance programme plan

Preventative maintenance

The preventative maintenance programme plan sets out all of the regularly scheduled system maintenance tasks, and includes which personnel are responsible, what their skill set must be, how to conduct the maintenance, which safe work method statements are to be followed, how to report the maintenance and so on. A subset of a typical maintenance programme is given in Table 6.1.

In addition, the maintenance programme plan specifies how work orders are approved and delivered, as well as how, where and by whom, maintenance logs are kept. Maintenance logs are generally kept using a computerised database. This makes it easier to perform the data reduction and analysis which may be necessary to find fault patterns or diagnose recurrent system issues.

Corrective maintenance plan

A corrective maintenance plan differs from preventative maintenance, as it does not involve a schedule. Instead, the plan involves specifying the maintenance activities that will take place in the event of certain circumstances and failures. A 'system condition' monitoring programme should be devised that allows problems to be identified, along with their root cause, and a system in place so that a trigger will be sent to the operator to schedule the relevant corrective maintenance activity.

Table 6.1 Example of a typical preventative maintenance programme

Activity	Personnel	SWMS	Frequency
Visual and mechanical inspection	Level 1 O&M Technician	SWMS X.Y.Z	Quarterly
Grounds maintenance	Grounds Maintenance Personnel	SWMS X.Y.Z	As necessary
String I–V curve testing	Level 2 O&M Technician (certified system installer/ electrician)	SWMS X.Y.Z	Annually
Cables, string/array protection, isolation/disconnection inspection testing	Level 2 O&M Technician (certified system installer/ electrician)	SWMS X.Y.Z	Annually
Transformer maintenance/relay calibration and battery testing	Level 3 O&M Technician (certified system installer/ electrician with HV ticket)	SWMS X.Y.Z	Annually
Infrared scans of modules, combiner boxes, switchgear and substation	Level 1 O&M Technician	SWMS X.Y.Z	Annually
Calibration, maintenance, and cleaning of on-site meteorological stations	Level 2 O&M Technician (certified system installer/ electrician)	SWMS X.Y.Z	Semi-annually, annually and fortnightly respectively
Inverter maintenance	Level 2 O&M Technician (certified system installer/ electrician)	SWMS X.Y.Z	As per manufacturer
Revenue meter verification	Level 2 O&M Technician (certified system installer/ electrician)	SWMS X.Y.Z	As per contract

Condition-based maintenance

A condition-based maintenance plan may be similar to the preventative maintenance plan, as it must specify all the maintenance activities that must be completed in order to maintain optimum performance and to adhere to system warranty requirements. The difference is that it requires planning of the monitoring system and corresponding software. This will involve determination of the different conditions that will result in the scheduling of certain maintenance activities.

Performance and availability plan

The performance and availability plan specifies the methodology for testing system performance. It is common for an independently calibrated weather station to be installed on-site and which includes one or many pyranometers for recording site-specific insolation data. Recorded weather data is used to calculate how much PV generated energy should have been created. The actual energy generated by the system is then compared with this calculated figure in order to arrive at a performance ratio. The performance ratio must be maintained at some contractually agreed level at all times (generally above 80%). The records of system availability show the runtime and downtime of the system. The operators should minimise downtime, ensuring the system is running about 99% of the time.

Maintenance procedures

There are many different maintenance procedures that must be performed over the lifetime of a solar farm. Examples of some of these activities are given in Figure 6.1. These procedures will include preventative maintenance, corrective maintenance and maybe even condition-based maintenance activities. When determining the extent of these maintenance procedures, it is very important to find the optimum balance between low cost and high plant reliability and performance.

Preventative maintenance

Preventative maintenance includes scheduled activities carried out to prevent issues occurring with the system equipment that may lead to a reduction in performance and loss of revenue. The extent and period of preventative maintenance activities will depend largely on the manufacturers' recommendations and the equipment warranty requirements. It will also depend on the specific environmental conditions, such as the type of vegetation on-site, the level of rainfall or snowfall, the dustiness or saltiness of the area etc. The climatic conditions and seasonable climatic variability are also discerning factors.

Preventative maintenance activities improve availability and performance, and hence maximise production but also prolong the life of the system. However, these activities increase the operating costs by incurring extra expenses and by loss of revenue during periods when the maintenance activities require the plant

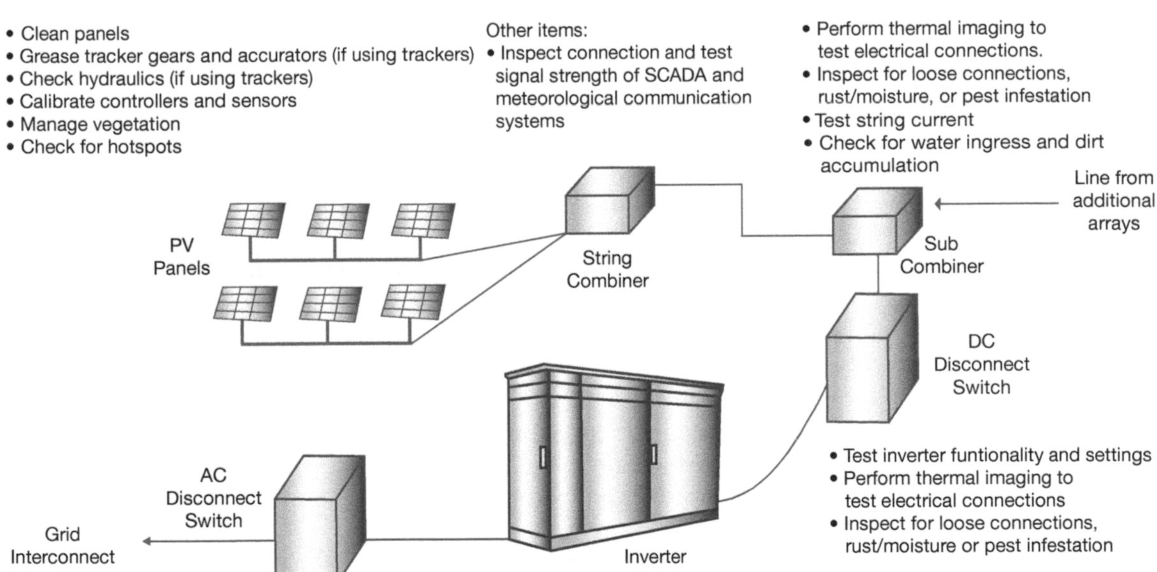

Figure 6.1 Some of the maintenance procedures required in order to operate a utility-scale solar farm. It is important that all maintenance procedures and scheduling are factored into the O&M budget.

Source: Scott Madden

to shut down. When designing the preventative maintenance schedule, it is important to find the optimum balance between the cost of these activities and the increased revenue from the increased yield over the system lifetime.

Preventative maintenance activities may include:

- Module cleaning
- Vegetation management
- Wildlife prevention
- Water drainage/erosion control
- Upkeep of data acquisition and monitoring systems
    - servicing electronics and sensors
- Upkeep of major components
    - inverter inspection and servicing
    - transformer inspection and servicing
    - checking module connection integrity
    - checking for hotspots
    - checking junction/combiner boxes for water ingress, dirt or dust accumulation and integrity of the connections within the boxes
- Upkeep of balance of system components
    - tracker maintenance
    - racking system maintenance
    - check all wiring
    - checking and servicing security system
    - checking signal strength and connection of communication systems.

Corrective maintenance

Corrective maintenance is performed in the event of failures within the system. When designing the corrective maintenance plan, it is important to consider how the problem will be diagnosed, how fast the response will be and how long the repair is expected to take. These parameters should be specified in the O&M contract and will depend on the site-specific conditions, as well as the availability and performance guarantees.

Again, while minimising the response and repair time will increase the yield of the system, it will also increase the cost of the repair. Therefore, it is important to balance the increased contractual costs of achieving shorter response times with the increase in revenue from increased yield.

Corrective maintenance is made up of:

- On-site monitoring/mitigation
- Critical reactive repair
    - address production losses
- Non-critical reactive repair
    - address production degradation
- Warranty enforcement (as needed).

While most corrective maintenance activities are typically related to inverter faults, they may also relate to a number of other components and include:

- tightening cable connections that have loosened;
- replacing blown fuses;
- repairing lightning damage;
- repairing equipment damaged by intruders or during module cleaning;
- rectifying SCADA faults;
- repairing mounting structure faults; and
- rectifying tracking system faults.

Condition-based maintenance

Condition-based maintenance employs real-time data to prioritise and optimise maintenance and resources. The plant operator will schedule O&M activities when deemed necessary based on the monitored data. While this approach may offer higher O&M efficiency, the cost of communication and monitoring systems required incur high upfront costs. In addition, any errors in communication or failures in the monitoring system may lead to poor maintenance scheduling, resulting in severe losses. While condition-based maintenance activities may be used instead of certain scheduled maintenance activities when it is more economic to do so, some manufacturers will require that their products are maintained according to a prescribed schedule in order for warranty claims to remain valid. Therefore, it is important to find the right balance between condition-based maintenance and preventative/scheduled maintenance activities (see Figure 6.2).

Condition-based maintenance may require the need to enforce warranty on components that have failed despite the recommended scheduled maintenance taking place.

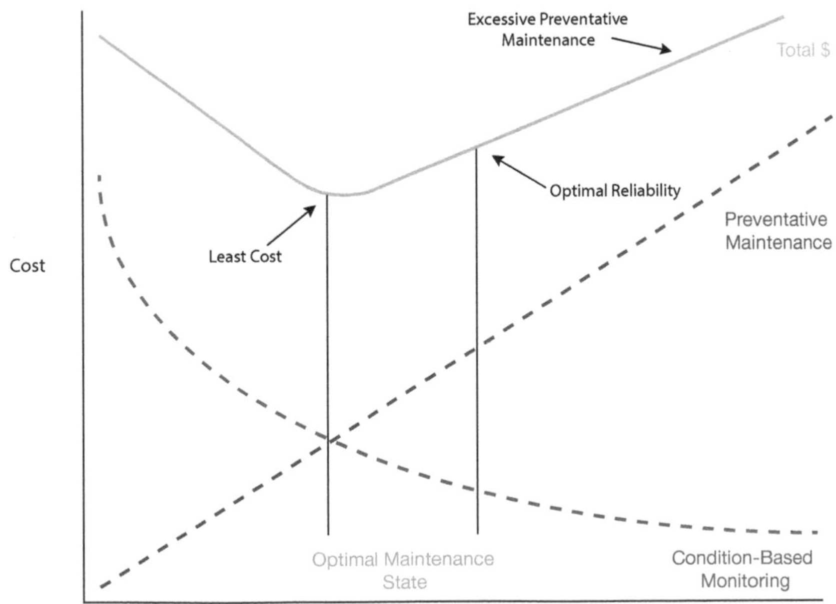

Figure 6.2 At excessively high levels of either preventative or condition-based maintenance, the O&M costs are very high. It is important to find the optimum balance between the two to find a solution which minimises O&M costs, but ensures adequate plant reliability. Note that this is just an example of what an optimal maintenance state may look like, in reality it will be more complex and will vary from project to project.

Source: S&C Electric Company

Personnel and responsibilities

The operations and maintenance procedures will require many different contractors with different levels of qualifications. The operations personnel manage the interaction with the grid and the sale of electricity. They also ensure that the plant is operating optimally, and may alert system maintenance personnel of any issues that they may discover. The system maintenance personnel are responsible for ensuring that all of the plant equipment is operating as it should, and must repair or replace equipment when necessary. The grounds crew are responsible for maintaining the site, including the grass, trees, debris and soil (erosion), as well as cleaning the modules.

Grounds crew

The grounds crew are responsible for maintaining the site on which the solar farm is constructed. This includes keeping pests and vegetation under control, but also includes cleaning the modules and managing erosion on and around the site. The crew must know what equipment to use in order to keep debris from damaging equipment, as well as know where and how to work in order to keep themselves and the equipment safe. It is important that all ground crew contractors have the relevant necessary qualifications in order to perform maintenance activities to a safe and reliable standard. The required qualifications will vary between countries and regions, but some recommendations are given in Table 6.2.

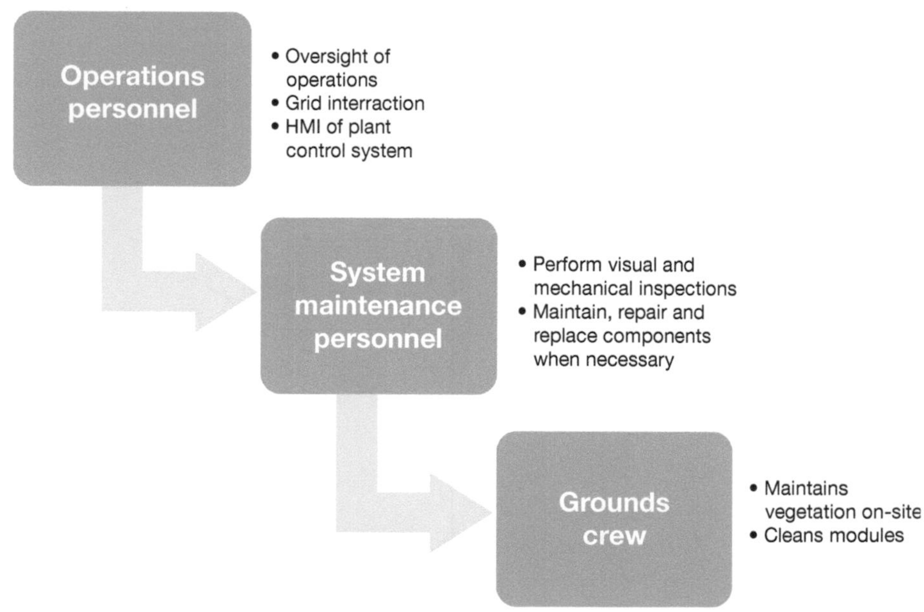

Figure 6.3 Hierarchy of O&M personnel and key responsibilities.

Table 6.2 Grounds crew contractor roles and qualifications

Contractor	Scope of work	Qualifications
Module cleaner	Cleaning PV array	• Training with the cleaning equipment which may include using a dust boom, brush trolley or crane • If a crane is used the array cleaner will need a licence to drive and operate the crane
Vegetation control	Trimming trees, cutting lawn, removing debris and applying herbicide	• Skilled in vegetation abatement and care • Training or licensed (if required) for herbicide application
Pest control	Removal and prevention of nesting vermin and insects	• Safety training in handling animals, debris and pesticides • May need licence for pesticide
Erosion control	Control and mitigate erosion and flood damage	• Trained and certified in erosion and sediment control • Previous experience

Source: NREL

System maintenance personnel

System maintenance personnel are responsible for performing visual and mechanical inspections, e.g. ensuring switched devices are fully operational etc., as well as more advanced tasks like string I–V curve testing and switchgear replacement. Some of these contractors must have even more advanced skills such as high-voltage equipment training and equipment calibration training for working on UPS controlled relays, relay calibration, transformer maintenance, etc. There is a wide range of system maintenance activities required throughout the lifetime of a solar farm, and it is important that all contractors have the necessary experience and qualification in order to perform these tasks safely and reliably. Some of the personnel required for system maintenance and the recommended qualifications are given in Table 6.3.

Operations personnel

Operations personnel will have to be familiar with the network requirements and the network's method for interacting with the system as well as the business plan of the site. To ensure that the plant meets the requirements on the network, the operator must be experienced with the HMI of the SCADA system, to enable him/her to input the required set points for power output control. The business plan may allow for very low operational oversight if the system is feeding the grid on a feed-in tariff arrangement or may require more regular oversight if the system is competing on a national electricity market.

Table 6.3 Qualifications and scope of work that is recommended of the contractors that may be required for system maintenance activities of a solar farm

Contractor	Scope of work	Qualifications
Structural engineer	Foundations and mounting frame inspection	• Engineering degree • Licence to practice in relevant state and country
Mechanic	Maintenance and repair/ replacement of tracking components	• Qualified mechanical contractor • Relevant experience
Electrician	Replacement and repair of electrical components	• Electrical contractor licence • PV installer certification • Experience in O&M of medium and high voltage electrical systems • 2–5 years of experience with PV systems • Training in special protective equipment and procedures
Network/IT/ SCADA	Internet/network repair, monitoring equipment repair	• Knowledge of specific monitoring devices (training by system supplier) • Knowledge of how monitoring system is connected through network connections or wireless or cellular modem • Knowledge of Modbus, DNP3 and other protocols • Knowledge of HMI operator interfaces • 2–5 years of experience • Knowledge of monitoring devices
Inspection	Diagnostic analysis; visual inspection, specific testing	• Diagnostic analysis • PV Installer certification • 2–5 years of experience
Inverter specialist	Inverter repair, upgrades	• Factory trained and certified to perform maintenance, diagnostics and repairs of inverters • 5+ years' experience
PV module/array specialist	Module repair	• Skills to operate, troubleshoot, maintain, and repair photovoltaic equipment • PV Installer certification • 2–5 years of experience

Monitoring

Weather monitoring

Weather monitoring is essential in order to evaluate and predict the PV plant's performance throughout its lifetime. The output data from the plant should be measured continually and adjusted with the actual weather parameters before comparing with the design values. This is because the output of PV systems constantly changes depending on the meteorological conditions like solar radiation and temperature. Weather data is also essential for analysing the weather pattern at the site in order to extrapolate future power generation trends (see Figure 6.4). Such analyses are valuable for the scheduling of O&M activities, setting control parameters for grid support and evaluating performance.

In order to monitor the weather accurately, a ground-based meteorological station should be installed on-site (see Figure 6.5). The weather station will include a range of meteorological instruments which may include but is not limited to those given in Table 6.4.

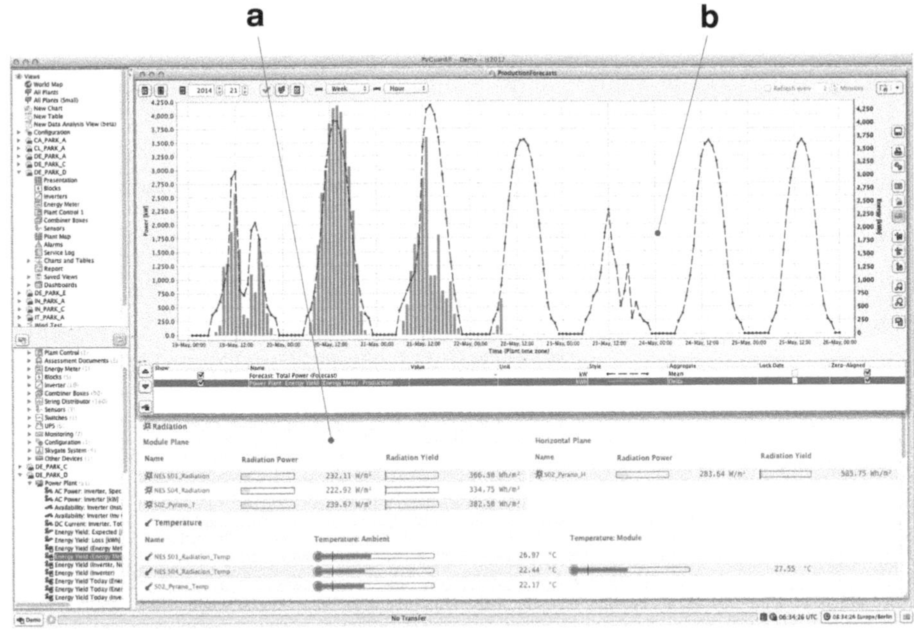

Figure 6.4 Energy production forecasts based on measured weather data. This PVGuard HMI shows (a) clear display of weather data and (b) output forecast to hourly resolution compared to actual production.

Source: skytron energy

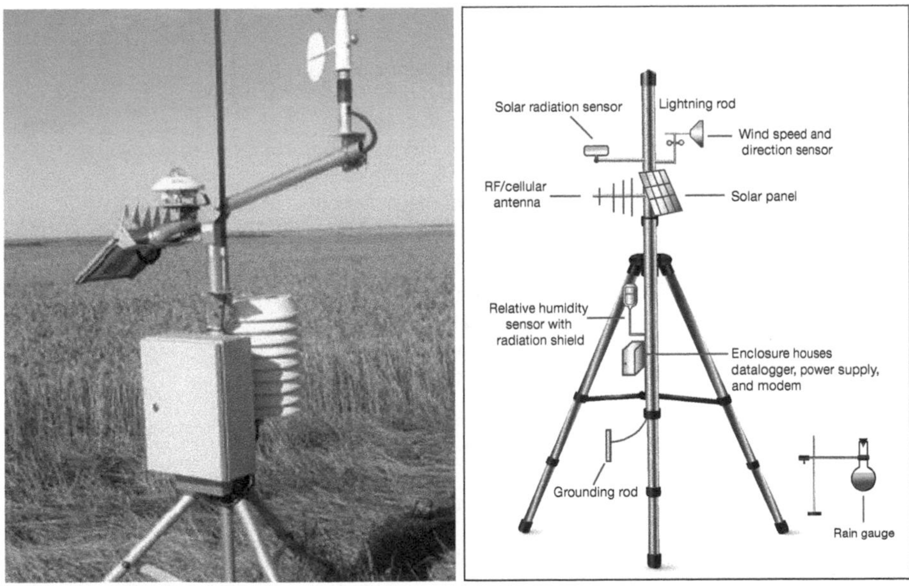

Figure 6.5 Automated weather monitoring station used for solar farm performance monitoring.

Source: First Solar

Table 6.4 Typical solar farm weather monitoring components

System Components	Function	Specifications
Pyranometer	Measures global horizontal irradiation (GHI, W/m^2) from a field of view of 180 degrees. Ventilation is essential	• Spectral range: typically 300–2,800 nm • Sensitivity (the lower, the better) • Operating temperature range and temperature dependence • Maximum solar irradiance • Response time
Pyrheliometer	Measures direct normal irradiance (DNI)	• Spectral range • Sensitivity (the lower, the better) • Operating temperature range and temperature dependence • Maximum solar irradiance • Response time
Anemometer and wind vane	Measures wind speed (m/s) and direction	• Wind speed range (m/s) • Starting threshold (m/s) • Wind direction range and delay distance • Operating temperature range • Accuracy (both speed and direction)
Relative humidity sensor	Measures relative humidity	• Sensing element material • Humidity range • Temperature operation range • Response time • Accuracy of measurements
Temperature probe	Measures ambient air temperature	• Temperature range • Accuracy of measurements
Surface temperature thermistor	Measures solar panel temperature	• Temperature range (needs high temperature tolerance) • Linearity • Accuracy of measurements
Ambient dust monitor	Monitors ambient dust (in g or mg) per m^3 of air	• Measurement principle • Concentration range (mg/m^3 or g/m^3) • Concentration sensitivity • Particle size range and particle size sensitivity • Humidity, temperature and pressure range • Accuracy of measurements • Flow rate (m/s)
Barometer	Measures pressure	• Sensor material • Digital or analogue output • Operating temperature • Measurement range • Accuracy of measurements
Rain gauge	Measures precipitation	• Sensor material • Sensitivity • Accuracy of measurements
Snow depth sensor	Measures snow depth	• Sensor material • Sensitivity • Accuracy of measurements

Source: Met One Instruments

System monitoring

System monitoring involves remotely monitoring all the major components of the PV plant to allow for easy troubleshooting and fast and cost-effective repairs. This ensures that the array operates at its peak performance for the majority of the project's lifetime. System monitoring is also crucial to enable effective power output controls. The plant is usually monitored via supervisory control and data acquisition (SCADA) systems (more information on SCADA is given in the next section, and has been discussed to an extent in Chapters 2 and 4). Monitoring may be installed at different levels depending on system requirements, each of which has its relative advantages and disadvantages as detailed in Table 6.5.

In order to achieve the required level of plant performance and availability, it is important that accurate data is recorded and interpreted at least on a daily,

Table 6.5 Comparison of different levels of monitoring within a PV power plant

Level of monitoring	Description	Advantages	Disadvantages
Inverter	• Shows inverter's status, adjustable parameters, and historic operating data • Offers foundational data for determining and maintaining basic system health	• Relatively low cost and high value • Recognising inverter problems through performance metrics can identify potential problems and enhance plant production when mitigated • Monitoring functionality is often embedded in inverters	• Level of resolution is limited to size and number of inverters used • Information gathering done either on-site or via remote link can be time consuming and labour intensive
Array	• Array monitoring collects information from DC circuits located in sections of a PV array	• Provides additional level of data without significant increased costs • Can isolate problem to a specific, though still large array section	• Limited benefit to cost trade off compared to inverter and string level monitoring • Fault panels will need to be identified by hand from within a group which increases labour costs
String	• Collects information at the string level	• Capable of determining the status of every string in the array • Root cause of problems are easy to identify from within a string	• Additional complexity, which requires special equipment and software to interpret and analyse data • Increased installation costs • Requires multiple communication devices which increases likelihood of failure and repair
Micro-inverter	• Allows monitoring at PV module level and leaves no need for DC cabling	• Efficiency benefits because they illuminate DC wire losses and need for parallel communication lines • Panel mismatch losses do not exist	• High costs • Relatively new technology which restricts financial backing from banks and other financiers

Source: Electric Power Research Institute, 2010

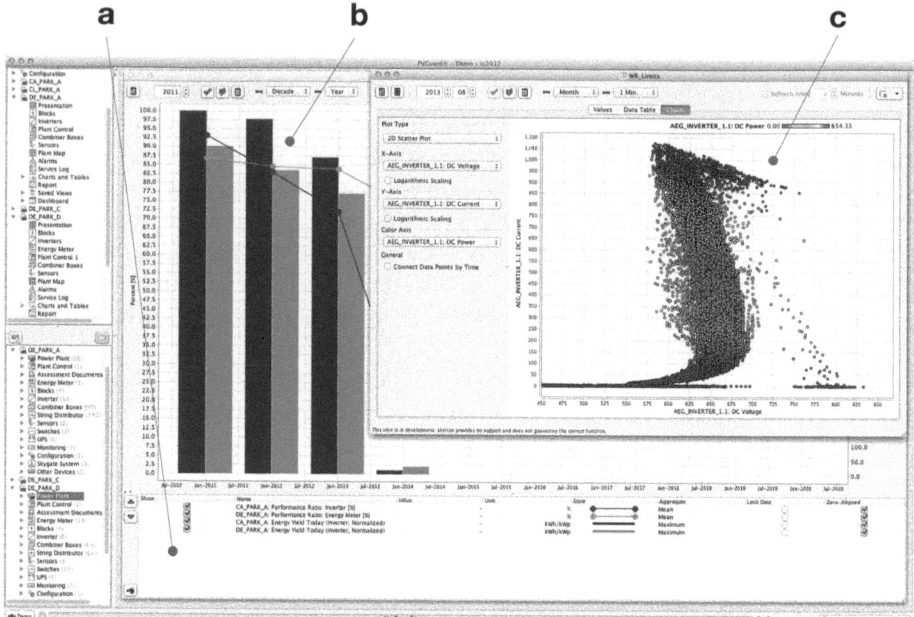

Figure 6.6 Data analysis of a utility-scale PV system showing (a) individual configuration of the way data is presented, (b) evaluation of historical data covering the entire plant operational life, (c) data analysis in a scatter plot.

Source: skytron energy

monthly and annual basis. This data should be stored and used to compare against the previous day's, month's or year-to-date production so that anomalies can be detected. Once the operation personnel are notified of an anomaly, they should be able to analyse the data to find the source of the problem so that the relevant system maintenance staff can be notified. The loss of production resulting from system faults should also be recorded, along with a record of the losses due to weather/insolation conditions that have resulted in lower than expected production.

The data is also used for the plant control system, to allow the system to interact with the grid so that the system adheres to requirements set by the grid operators, and provides grid support where possible. More information in given below in the section on power output controls.

In order to have accurate data to monitor, it is important that all monitoring equipment is of high quality and in good working order. The hardware and software used to monitor the system should be maintained, serviced and calibrated regularly by the relevant service providers. It is also important that the data is backed up regularly and modelled by a variety of sources to ensure reliability.

Fault detection

Faults and failures in the system can range from being very minor, such as temporary ground fault due to rain, to serious issues, such as output from the plant failing to synchronise with the grid in terms of voltage, PF or frequency. Either way, it is important that operations personnel are notified in some way of the fault or failure so that appropriate action can be taken based on the severity of issue. For example, if a blown string fuse is detected, it would likely be more economic to have it replaced during scheduled maintenance rather than sending maintenance personnel out to the site to fix it immediately. However, if a tripped breaker is detected in the AC switchgear, it is likely that immediate response is necessary, as this would affect a large portion of the PV plant, and may reduce the yield significantly.

String monitoring remains the convention, but as systems become increasingly larger, the level and type of monitoring may change over time. String faults will be detected by string monitoring which are basically current transformers, and are typically built into string combiner boxes. DC isolation is also built into string combiner boxes, and often has the ability to remote trip as signalled by the SCADA system.

Monitoring provided at the inverter level will provide the majority of fault detection capability. These days most utility-scale inverters are embedded with monitoring capability which includes:

- Monitoring voltage, current, power and energy (daily, monthly, lifetime) for the sub-array that they are connected to.
- Keeping a fault log, including residual current levels and ground fault detection.
- Providing active inverter comparison (where inverter output is compared in real time against predefined tolerances).
- Showing communication (comms) status (i.e. any comms dropouts, length of time without comms, etc.).
- Showing inverter attenuation level based on SCADA or network control.

Inverter monitoring also records a myriad of grid-side status and fault information including line voltage, frequency, power factor, insulation resistance. It may also record more detailed grid parameters such as vector shift, rate of change of frequency, and total harmonic distortion.

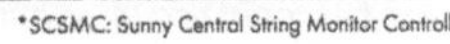

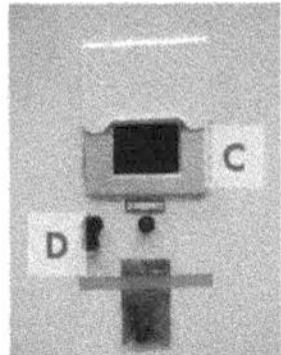

A. Splice box
B. FO-Converter
C. Display
D. Service Ethernet
E. Switch-disconnect
F. WebBox
G. SC20CONT
H. SCSMC*

Figure 6.7 Inside an SMA Sunny Central CP inverter with string monitoring equipment.

Source: SMA Solar Technology AG

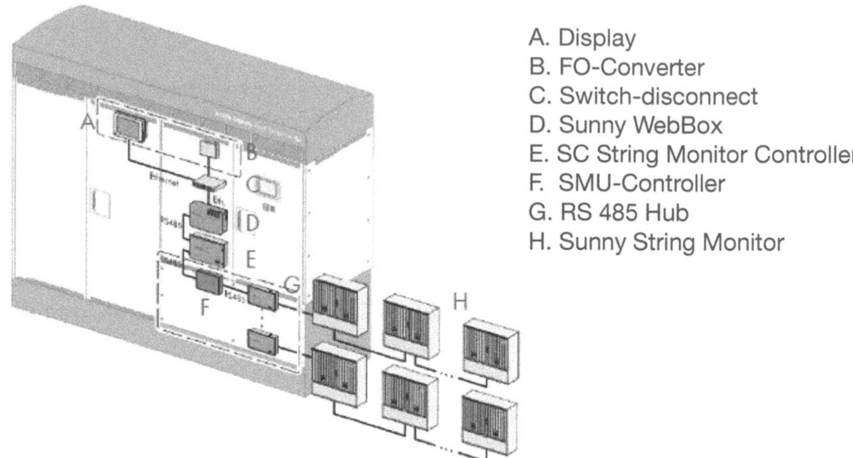

Figure 6.8 Example of SMA Sunny Central CP string monitoring configuration.

Source: SMA Solar Technology AG

The operational personnel may be notified of a fault or failure in the system via either of the following ways:

- An alarm or error report sent from the monitoring system
- Routine inspection conducted by maintenance personnel
- Performance analysis of the power plant
- An alarm or error report sent from a third party such as a network service provider.

Supervisory control and data acquisition (SCADA) systems

Solar farms require a range of data acquisition and control equipment. The system, which combines this equipment and the communication network that connects them all, is called the SCADA system. The SCADA system gathers field information from the weather station, string monitoring, inverters, step-up transformers, switch gear, site security systems etc. The SCADA system will parse the information in real time to ensure the system is operating as expected and will trigger alarms if it is not. The SCADA system may be able to respond to information with automatic controls or may need operator input using a human–machine interface (HMI).

The basic framework comprising the SCADA system is:

- Field devices include transducers (i.e. CTs, VTs, weather station equipment, etc.) and operational control gear (i.e. contactors, motorised circuit protection, relay state, etc.).
- Remote telemetry units (RTUs) and programmable logic controllers (PLCs) connect the field devices to the communication system for data collection, aggregation, analysis and response.
- The communication architecture consists of the infrastructure and protocol for data packaging and transmission.
- SCADA host platform takes the data from RTUs and PLCs, stores and processes it to show graphical displays, trends, trigger alarms and generate reports.
- The human–machine interface (HMI) most often uses a graphical user interface (GUI) set up to connect human operators with process monitoring and control.

Field devices

Field devices are the instruments that are installed at each component to measure their performance. These devices include sensors and transducers used to measure the operating parameters of the system, and actuators that implement the control signals.

Field devices may include:

- Current transformers (sometimes called current transducers) are sensors that measure alternating current and are very useful for string monitoring, sub-array monitoring, inverter AC output monitoring, and step-up transformer monitoring.
- Voltage monitoring instruments (sometimes referred to as voltage transformers or transducers or VTs) for inverter DC input, inverter AC output and step-up transformer monitoring.
- Internal inverter state monitoring, which provides watchdog control for all inverter parameters including connection faults, grid faults, earth faults, and other Boolean faults.
- Weather station devices which are described above.
- Operational control gear such as contactors, relays, motors.

The SCADA system becomes more complex if sun-path tracking devices, energy storage systems, fixed grid restrictions (i.e. connection to a local diesel grid) or grid-support functionality is included in the system. This requires more field devices and more complex monitoring hardware and software.

Programmable logic controllers and remote telemetry units

PLCs and RTUs are used to connect the field devices to the communication system for data collection, aggregation, analysis and response.

An RTU is a microprocessor controlled electronic device that transmits data from the field to the comms system using well-developed communication interfaces or telemetry, but may or may not have programmable logic control.

PLCs are very similar to RTUs in their functionalities, except that they are capable of providing controlled logic to a circuit, but don't have communication interfaces or data logging capability. Therefore, these devices provide much more flexibility and functionality when paired together. However, these days RTUs are built with control programming and PLCs are built with a level of telemetry and data logging functionality, so that they almost operate the same as each other. These devices can be reprogrammed, debugged and fixed in the field itself. This enables them to execute simple logical processes without involving the server station.

Communications network

The data from RTUs and PLCs to the control centre and commands from the control centre to RTUs and PLCs is transmitted over the communication network. This may be done by creating a local and distributed network, by aggregating local networks for centralised control across a wide area network or by creating a horizontal communication topology where all systems equipment are independently connected and centrally controlled. The actual communication infrastructure may use powerline communications, analogue or digital control cables, Ethernet or fibre optic communication cable or use any number of wireless protocols.

SCADA host platform

The SCADA host platform consists of software that continuously retrieves, stores and processes the data from the RTUs and PLCs into interactive forms like tables and graphs; and generates necessary alarms and reports. The server station can also offer automated and/or manual modes of control of different components of the system by sending commands to the RTUs or PLCs.

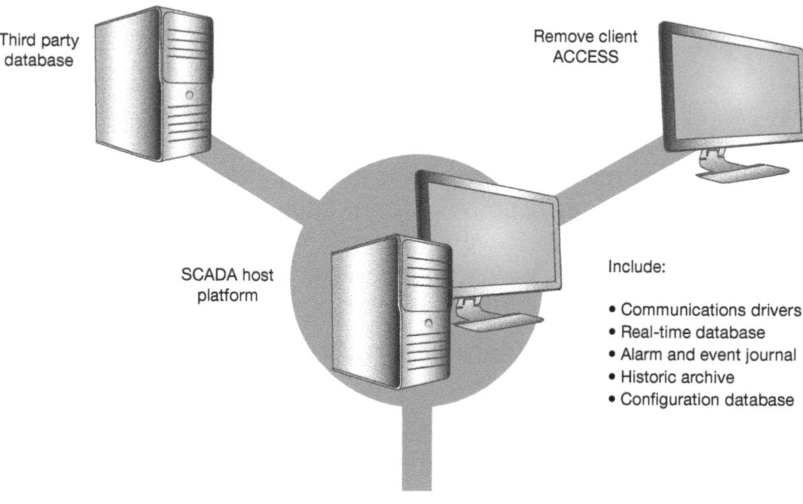

Figure 6.9 Interaction of the SCADA host platform with data from the field, third party database and the HMI.

The platform backs up all the information retrieved from the system including information on performance, production and maintenance, status reports, etc. in an archiving server which ensures that the data is available in case the system crashes or becomes corrupted. The historic archive is usually backed up on both local and remote workstations for additional redundancy. The SCADA host platform also includes a real-time database server where users or system administrators can query the historical data in order to statisticise or analyse the performance of PV plant.

Human–machine interface

An HMI presents the data stored in the host computers to the human operator in an understandable and comprehensible form. This may include trending, diagnostic or management information and detailed schematics and graphics representing the current status of equipment under its control.

HMI also allows operations personnel to interact with the plant to allow for grid support functionality such as voltage regulation, reactive power control, active power curtailment, ramp-rate and frequency control. It can also be used in other ways such as for fault isolation, routine maintenance and electricity market participation.

A screen shot from a SCADA HMI is given in Figure 6.11 showing information on the current state of individual inverters in a PV plant along with their real-time output power, performance index and alarm status. The pictorial

Figure 6.10 Operations personnel monitoring a utility-scale solar system using skytron HMI.

Source: skytron energy – Operations Center

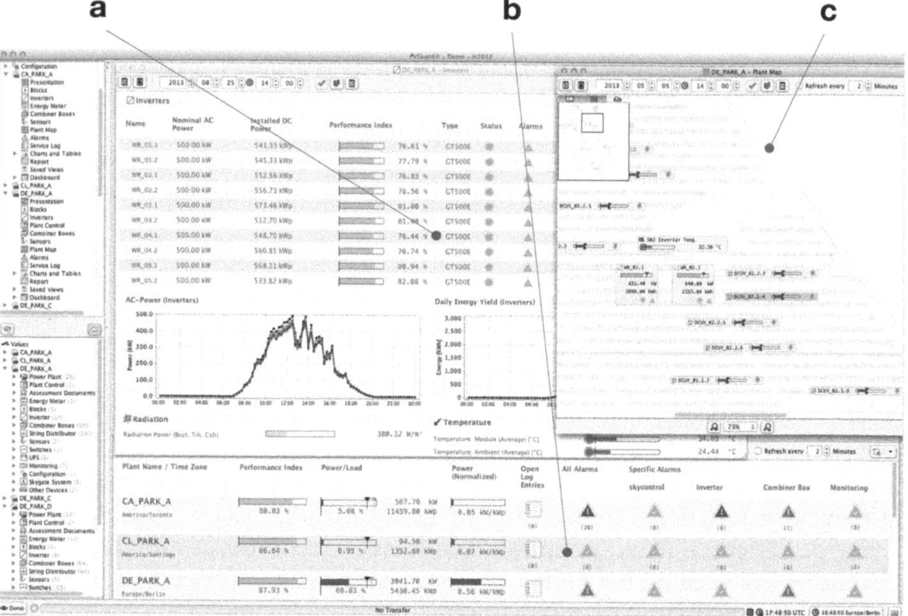

Figure 6.11 SCADA HMI screen shot showing (a) inverter alarms and status; (b) overview of alarms from all plants and (c) alarms depicted in site map for geographic orientation.

Source: skytron energy

representation of time-stamped generation from inverters helps to quickly compare and analyse their performances. On the left navigation pane of the screen, there is a nested list of a large number of options that define different operating parameters of plant equipment, for example, DC current string. This shows how the status of the whole PV plant can be accessed very easily through the HMI of a SCADA system.

Power output controls/plant controller

In order for a variable output generator like a utility-scale solar plant to be able to be connected to the grid it must address the specific requirements of the local grid. This is done using a plant controller incorporated into the SCADA system. The plant operator can monitor and send commands to the controller via the SCADA HMI. The plant controller then uses plant-level logic and closed-loop control systems to command the inverter in real time to achieve fast and reliable regulation. The plant controller should provide the following functions:

- Dynamic voltage and/or power factor regulation of the solar plant at the POI (point of interconnection).
- Real power output curtailment of the solar plant when required, so that it remains within the limits specified by the utility operator.
- Ramp-rate controls to ensure that the plant output does not ramp up or down faster than a specified ramp-rate limit (rate of ramp down may require the use of energy storage).

- Frequency control to lower plant output in case of over-frequency situation or increase plant output (if possible) in case of under-frequency.
- Start-up and shut-down control.

For a solar farm to perform these power output controls, it often means that the output is curtailed, leading to loss of revenue. However, some electricity markets around the world pay generators for providing grid-support functions through frequency control ancillary services (FCAS). Revenue from this market may compensate for the loss of output from curtailment. For example, a 1.2 MW solar farm developed in Hawaii, was restricted by the local utility to generate only 600 kW to ensure the reliability of weak grid (4 MW) on the island. This resulted in significant loss in revenue for the project. However, once the value in grid management services was identified, the plant was allowed to operate at full capacity to provide ramp-rate control, frequency response and power quality management.

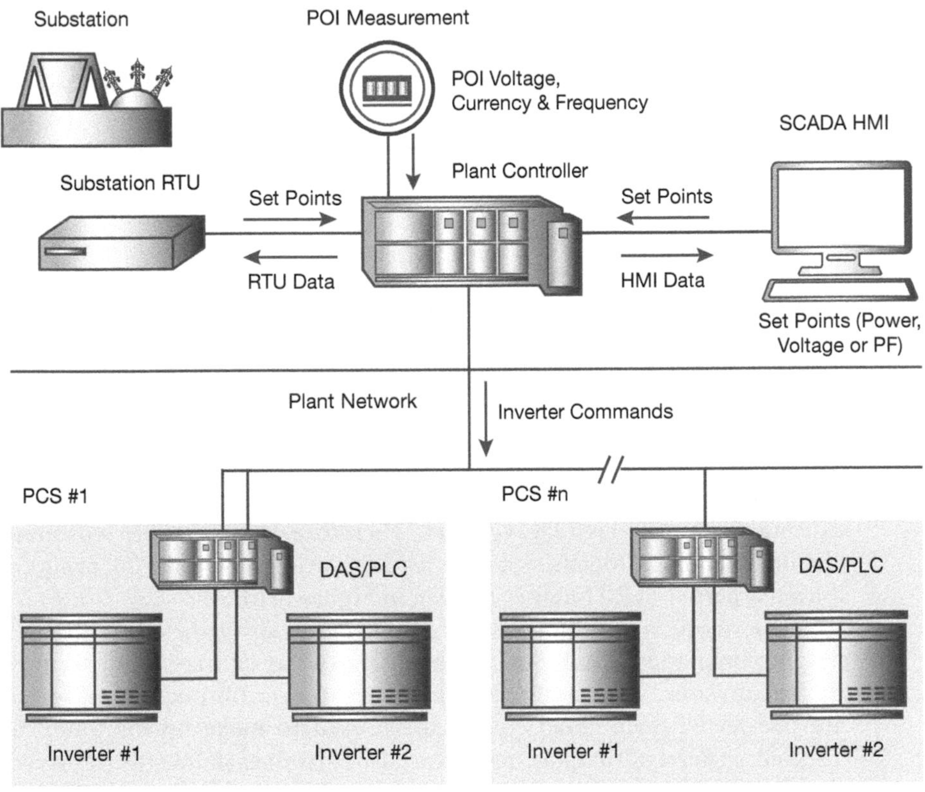

Legend				
DAS	Data Acquisition System		HMI	Human Machine Interface
PCS	Power Conversion Station		PLC	Programmable Logic Controller
POI	Point of Interconnection		RTU	Remote Terminal Unit
SCADA	Supervisory Control & Data Acquisition System			

Figure 6.12 Plant control system and interfaces with other components.

Voltage and reactive power regulation

Voltage regulation on the transmission level of the utility grid can be achieved using synchronous generators, as well as deploying static and dynamic devices where necessary along the transmission network. However, this can be costly for utility operators, so it is important and mandatory for variable output plants like solar farms to be able to play their part in voltage regulation and reactive power control. This can be achieved using the inverters' voltage regulation capabilities, operated through the plant control system.

The plant control system should include some sort of automatic voltage regulation (AVR) functionality, of which the plant operator can select one of three modes:

1 Voltage control: the controller regulates the reactive power produced by the plant to ensure the voltage remains at the required set point at the POI.
2 Power-factor mode: the controller commands the inverters to meet the power-factor set point by managing their reactive power output.
3 VAR mode: The controller ensures that the VARs at the POI are within the required set point by commanding the inverters to inject reactive power during voltage sags and absorb reactive power during voltage rise.

The plant operator can change the set points depending on the utility requirements during a given period of time using the SCADA HMI for AVR.

Power curtailment

Depending on local demand and grid requirements, there will be periods when the PV plant will need to curtail the power output to ensure that there is no excess active power on the transmission network. This set point can also be adjusted as required by the plant operator via the HMI. When the controller detects that the reactive power at the POI is above this set point it calculates and commands individual inverters to curtail their output. However, inverters are only capable of reducing their active power output to a certain level without causing excessively high DC voltages. Therefore, the controller will shut down certain inverters all together when necessary. An example of PV plant curtailment during a period of 20 minutes is given in Figure 6.13.

Furthermore, during periods of variable cloud cover the controller will send unique commands to each inverter based on the specific conditions experienced by that inverter. This means that the excess power output from inverters that are unaffected from cloud cover can be used to make up for those that are affected, instead of ramping down the output from each inverter. This increases the overall yield from the solar farm, and is illustrated in Figure 6.14.

Ramp-rate control

The output of utility-scale PV systems is typically highly variable at short time periods due to changes in cloud cover. This unpredictability in power output (ramps) can strain the electrical grid by making it more difficult to match power

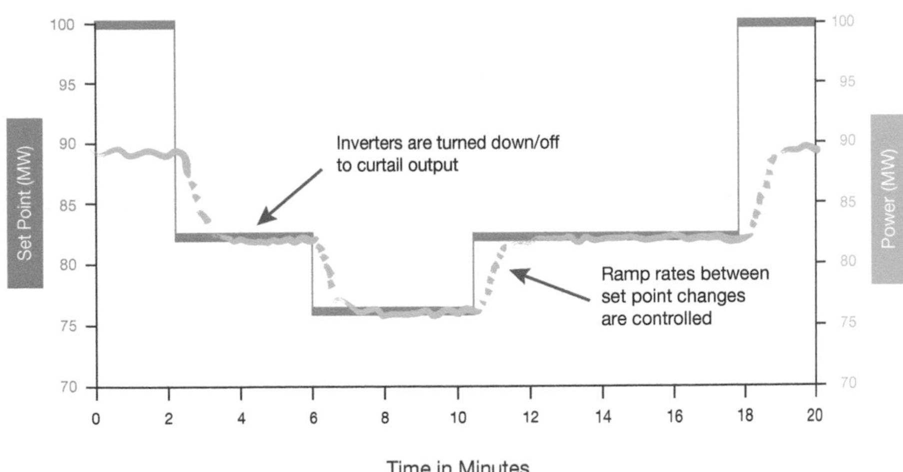

Figure 6.13 A plant operating at 90 MW. At the beginning of the period the active power set point is above the operating point so the plant is not curtailed. After two minutes the plant must curtail its output to 82.5 MW, and the controller responds as expected to reduce the output of the plant at the POI. The set point is then further reduced to 75 MW, and then gradually back up to 100 MW. During each change in active power set point, the controller responds by commanding the inverters to ramp up or down, or switch on and off in a gradual manor to ensure that the ramp-rate limit is maintained.

Source: First Solar, 2015

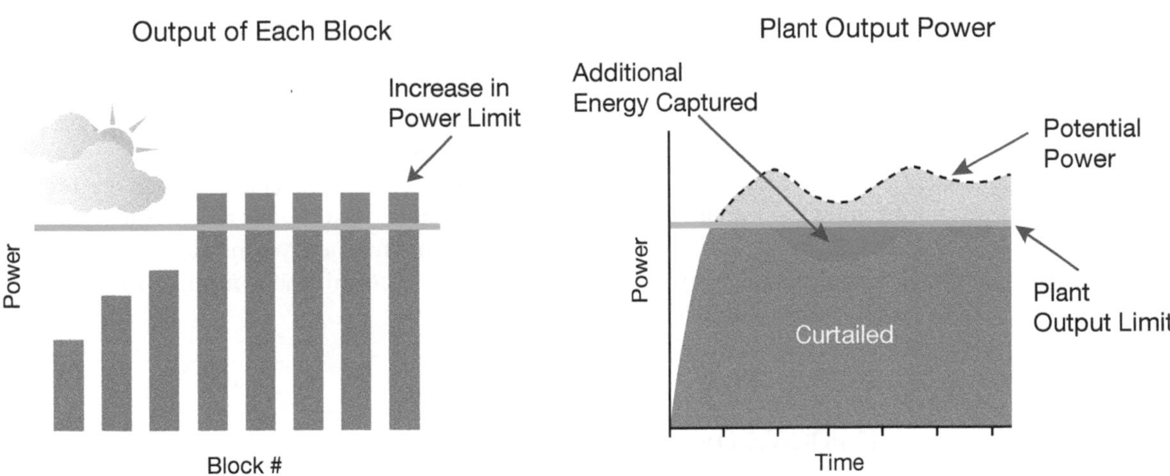

Figure 6.14 (a) This shows that for a given period of time, each inverter (block) may have a different output. Those that are above the output limit can have their allowable output increased to make up for the inverters generating below the output limit. (b) This shows how the inverters that are unaffected from cloud cover can make up for those that are affected, so that more of the available capacity is utilised when the plant is curtailed.

Source: First Solar, 2015

delivery with electricity demand meaning that the utility must maintain spinning reserve capacity (back-up supply) to make up for sudden drops in power supply. For this reason, the grid operator will often specify a maximum ramp rate for solar farms in terms of megawatts per minute.

There are several approaches that can be used to minimise the impact of variable generation from PV plants. Some of the approaches include:

- Reactive power injection: voltage control can be achieved through reactive power injection from PV inverters using the plant controller as explained above.
- Geographical dispersion: if the solar farm is more geographically dispersed, different parts of the array will experience ramps at different periods. This means that the overall variability of the plant is reduced because it is averaged out over a larger area. Note that this is more of a planning/design consideration.
- Solar forecasting: reliable solar forecasting and management tools can be used to provide forewarning that output is likely to vary. In response to these warnings, the plant operator can curtail the plant output prior to the fluctuation and grid operators can prepare spinning reserve capacity.
- Storage: energy storage systems can be used to absorb and deliver power to the grid to limit fluctuations in power output. Storage can also be used to provide several other services such as peak shaving, load shifting, and back-up power during outage.

Frequency control

When there is insufficient grid power to meet the demand on the utility grid, grid frequency decreases; when there is a surplus of generation relative to demand, the frequency increases. In other words, frequency is an indication of the imbalance between supply and demand on the utility grid, and can be used to trigger active power control by generators.

The intermittency of solar generation can cause challenges in regulating frequency. Therefore, the plant control system must be able to provide frequency control in the case of above or below normal frequency conditions. This is done by commanding the inverters to increase or curtail their output to help match grid supply with demand.

Bibliography

Energy Storage Association. *Solar PV-Storage: Lanai Sustainability Research Dynamic Power Resource (DPR®) Energy Storage*. n.d. (accessed 28 February 2016).

First solar. *Grid Integration*. 2015.

GSES. 'Chapter 17 – System Commissioning'. In *Grid-Connected PV Systems: Design and Installation*. 2015.

ISF. 'Utility-scale Solar Guide'. 2015.

Matthew Lave, Jan Kleissl, Abraham Ellis and Felipe Mejia. *Simulated PV Power Plant Variability: Impact of Utility-imposed Ramp Limitations in Puerto Rico*. 2013.

Met One Instruments. 'Weather Station for Solar Farm Monitoring'. n.d.

NREL. *Best Practices in PV System Operations and Maintenance* . Solar Access to Public Capital (SAPC) Working Group, 2015.

Appendix: Resources

Guidelines and regulations

Guidelines and regulations play a key role in the successful design, construction and operation of solar farms. While there are no standards that are explicitly intended for utility-scale solar, there are many general standards that are applied to utility-scale systems. There are also various guidelines and regulations set and applied by different countries, and across different states and regions.

Guidelines associated with solar farm development

While there are limited comprehensive guidelines available for the planning and development of solar farms, there are various guidelines available for different countries corresponding to different aspects of development. Some of these include:

Utility-Scale Solar Photovoltaic Power Plants: A Project Developer's Guide (International Finance Corporation, 2015), www.ifc.org/wps/wcm/connect/f05d3e00498e0841bb6fbbe54d141794/IFC+Solar+Report_Web+_08+05.pdf?MOD=AJPERES This guidebook provides good information on all stages of solar farm development including site identification, plant design, energy yield, permits/licences, contractual arrangements, and financing. It is aimed at project developers entering the market, and meant as a reference source for contractors, investors, government decision makers, and other stakeholders working on PV projects.

Planning Practice Guidance for Renewable and Low Carbon Energy – Department for Communities and Local Government (updated 18/06/2015), http://planningguidance.communities.gov.uk/blog/guidance/renewable-and-low-carbon-energy/ These guidelines provide advice from the UK government on the planning issues associated with the development of renewable energy projects, and are expected to be followed unless there are adequate reasons not to.

Guidelines for Landscape and Visual Impact Assessment (Landscape Institute, 2013), www.landscapeinstitute.org/knowledge/GLVIA.php Guidelines, such as these, aim to address the visual aesthetics of these installations and help practitioners to achieve quality and consistency in their approach to land and visual impact assessments in the UK.

UK Planning Guidance for the Development of Large Scale Ground-Mounted Solar PV Systems (BRE National Solar Centre, 2013), www.bre.co.uk/filelibrary/pdf/other_pdfs/KN5524_Planning_Guidance_reduced.pdf This document provides best practice planning guidance for how large ground-mounted arrays are developed in the UK. It describes planning considerations and requirements.

Biodiversity Guidance for Solar Developments (BRE National Solar Centre, eds G. E. Parker and L. Greene, 2014), www.bre.co.uk/filelibrary/pdf/Brochures/NSC-Biodiversity-Guidance.pdf These guidelines provide information to solar farm developers in the UK on how they can support biodiversity on solar farms. It presents a broad range of options for biodiversity enhancement and management, and illustrates best practice through a series of case studies.

Agricultural Good Practice Guidance for Solar Farms (BRE National Solar Centre, ed. J. Scurlock, 2014), www.bre.co.uk/filelibrary/nsc/Documents%20Library/NSC%20Publications/NSC_-Guid_Agricultural-good-practice-for-SFs_0914.pdf These guidelines detail the principles of good practice for the management of small livestock in solar farms established on agricultural land, derelict/marginal land and previously developed land in the UK.

Technical Guideline: Generating Plants Connected to the Medium-Voltage Network (BDEW, 2008), www.bdew.de/internet.nsf/id/A2A0475F2FAE8F44C12578300047C92F/$file/BDEW_RL_EA-am-MS-Netz_engl.pdf This guideline summarises the essential aspects that must be taken into consideration for the connection of generating plants to the medium-voltage network to assist in the planning and decision making of network operators and installers in Germany.

Land-Use Requirements for Solar Power Plants in the United States (NREL, 2013), www.nrel.gov/docs/fy13osti/56290.pdf This report provides data and analysis of the land use associated with US utility-scale PV and concentrating solar power facilities, in response to concerns regarding large-scale deployment of solar energy and its potentially significant land use.

Planning and Zoning for Solar Energy Systems (American Planning Association, 2011), www.planning.org/pas/infopackets/open/pdf/30intro.pdf This Essential Info Packet provides a number of articles and guidebooks for various states and counties across the US to provide information to planners and developers on the current state of solar and help them plan for solar in these areas.

Establishing the Social Licence to Operate Large Scale Solar Facilities in Australia: Insights from Social Research for Industry (ARENA), http://apo.org.au/files/Resource/ipsos-arena_solarreport.pdf This document describes best practice principles for establishing social licence to operate (gaining community acceptance) utility-scale solar in Australia. It is based on research with the Australian general public, as well as specific research in communities living in close proximity to planned or established solar farms.

International Standards

Institute of Electrical and Electronics Engineers – IEEE Series

The Institute of Electrical and Electronics Engineers in the USA developed international regulation: the *IEEE family*. The following standards refer to grid interconnection:

- IEEE 929 – IEEE Recommended Practice for Utility Interface of Photovoltaic (PV) Systems
- IEEE 1547 series Standard for Interconnecting Distributed Resources with Electric Power Systems (includes relevant information for utility-scale systems)
- IEEE P57.159/D6 – Draft Guide on Transformers for Application in Distributed Photovoltaic (DPV) Power Generation Systems (includes utility-scale systems)
- ANSI/IEEE C37.2–2008 Standard Electrical Power System Device Function Numbers, Acronyms and Contact Designations.

International Electrotechnical Commission – IEC/TS Series

The IEC/TS is the world's leading organisation for the preparation and publication of international standards for all electrical, electronic and related technologies. IEC/TS Technical Committee 82 prepares International Standards related to all aspects of installing a grid-connected solar PV system and is used by many countries around the world. Some of these standards include:

General

- IEC/TS 60904 Series Photovoltaic devices
  - *Part 1* Measurement of photovoltaic current-voltage characteristics
  - *Part 2* Requirements for photovoltaic reference devices
  - *Part 3* Measurement principles for terrestrial photovoltaic (PV) solar devices with reference to spectral irradiance data
  - *Part 4* Reference solar devices – Procedures for establishing calibration traceability
  - *Part 5* Determination of the equivalent cell temperature (ECT) of photovoltaic (PV) devices by the open-circuit voltage method
  - *Part 7* Computation of the spectral mismatch correction for measurements of photovoltaic devices
  - *Part 8* Measurement of spectral responsivity of a photovoltaic (PV) device
  - *Part 9* Solar simulator performance requirements
  - *Part 10* Methods of linearity measurement
- IEC/TS 61215 Crystalline silicon terrestrial photovoltaic (PV) modules – Design qualification and type approval
- IEC/TS 61646 Thin-film terrestrial photovoltaic (PV) modules – Design qualification and type approval
- IEC/TS 62446 Grid connected photovoltaic systems – Minimum requirements for system documentation, commissioning tests and inspection
- IEC/TS 62941 Terrestrial photovoltaic (PV) modules – Guideline for increased confidence in PV module design qualification and type approval
- IEC/TS 61727 Photovoltaic (PV) systems – Characteristic of the utility interface
- IEC/TS 60891 Procedures for temperature and irradiance corrections to measured I–V characteristics of crystalline silicon photovoltaic (PV) devices

- IEC/TS 62040 series Uninterruptible power systems
  - *Part 1* General and safety requirements for UPS
  - *Part 2* Electromagnetic compatibility (EMC) requirements
  - *Part 3* Method of specifying the performance and test requirements
  - *Part 4* Environmental aspects – Requirements and reporting.

Testing

- IEC/TS 61853 Photovoltaic (PV) module performance testing and energy rating
  - *Part 1* Irradiance and temperature performance measurements and power rating
  - *Part 2* Angle of incidence effect on photovoltaic modules
- IEC/TS 62116 testing procedure of islanding prevention measures for power conditioners used in grid-connected photovoltaic power generation systems
- Electrical safety of installation
- IEC/TS 61345 UV test for photovoltaic (PV) modules
- IEC/TS 61829 On-site measurement of current-voltage characteristics
- IEC/TS 62116 Utility-interconnected photovoltaic inverters – Test procedure of islanding prevention measures
- IEC/TS 62910 Utility-interconnected photovoltaic inverters – Test procedure for low voltage ride-through measurements
- IEC/TS 62804 Photovoltaic (PV) modules – Test methods for the detection of potential-induced degradation
- IEC/TS 62852 Connectors for DC-application in photovoltaic systems – Safety requirements and tests

Balance of system components

- IEC/TS 62093 Balance of system components for photovoltaic systems – Design qualification natural environments
- IEC/TS 61683 PV systems – Power conditioners – Procedure for measuring efficiency
- IEC/TS 62093 Balance-of-system components for photovoltaic systems – Design qualification natural environments
- IEC/TS 60529 Degrees of protection provided by enclosures (IP Code)
- IEC/TS 62727 Photovoltaic systems – Specification for solar trackers
- IEC/TS 62817 Photovoltaic systems – Design qualification of solar trackers
- IEC/TS 62790 Junction boxes for photovoltaic modules – Safety requirements and tests
- IEC/TS 62894 Photovoltaic inverters – Data sheet and name plate

Operation

- IEC/TS 61724 Photovoltaic system performance monitoring – Guidelines for measurement, data exchange and analysis
- IEC/TS 61725 Analytical expression for daily solar profiles

- IEC/TS 61936 Series Power installations exceeding 1 kV a.c.
- IEC/TS 61727 Photovoltaic (PV) systems – Characteristics of the utility interface.

Safety

- IEC/TS 62109 Series Safety of power converters for use in photovoltaic power systems
- IEC/TS 60950 Safety of information technology equipment
- IEC/TS 62109 Safety of power converters for use in photovoltaic power systems
  - *Part 1* General requirements
  - *Part 2* Particular requirements for inverters
- IEC/TS 61557 series Electrical safety in low voltage distribution systems up to 1,000 V a.c. and 1 500 V d.c.
- IEC/TS 62548 Installation and safety requirements for photovoltaic (PV) generators
- IEC/TS 61730 PV module safety qualification
  - *Part 1* requirements for construction
  - *Part 2* requirements for testing.

Cabling

- IEC/TS 60502 for cables between 1 kV and 36 kV
- IEC/TS 60364 series for LV cabling
- IEC/TS 60840 for cables rated for voltages above 30 kV and up to 150 kV
- IEC/TS 61386 conduit systems for cable management

IEC/TS standards are commonly applied in the world by different countries including Britain, Germany, France, Italy, Brazil and China. It should be pointed out that since each country has specific supply characteristics, the IEC/TS standards applied differ from one country to another.

International Organization for Standardization – ISO Series

ISO is an independent, non-governmental international organisation that develops market relevant International Standards to ensure that products and services are safe, reliable and of good quality.

General project related standards

- ISO 9000 Quality Management
- ISO 45001 Occupational Health and Safety management systems – Requirements (due for release in October 2016)
- ISO 14001 Environment Management
- ISO 31000 Risk Management

PV related standards

- ISO 9845 Solar energy – Reference solar spectral irradiance at the ground at different receiving conditions
    - *Part 1* Direct normal and hemispherical solar irradiance for air mass 1.5
- ISO 9060 Solar energy – Specification and classification of instruments for measuring hemispherical solar and direct solar radiation
- ISO 9846 Solar energy – Calibration of a pyranometer using a pyrheliometer.

National codes and product standards

The National Electrical Code (NEC)

NEC is the benchmark for safe electrical design, installation, and inspection to protect people and property from electrical hazards in the US.

- NEC 690 Solar Photovoltaic (PV) Systems
- NEC 250 Grounding and bonding.

The European Standards (EN Series)

These standards are mainly managed by European Committee for Standardization (CEN), European Committee for Electrotechnical Standardization (CENELEC) and European Telecommunications Standards Institute (ETSI). The EN Series associated with grid-connected PV systems include:

- EN 50380 Data sheet and nameplate information for non-concentrating photovoltaic modules
- EN 61215 Crystalline silicon terrestrial photovoltaic (PV)
- EN 61646 Thin-film terrestrial photovoltaic (PV) modules – Design qualification and type approval
- EN 50461 Solar cells – Datasheet Information and product data for crystalline silicon solar cells
- EN 50116 Information technology equipment – routine electrical safety testing in production.

As conditions vary by countries, a number of national standards are developed based on EN series, such as BS EN which can be applied to British as well as many European countries, and DIN EN, which is a combination of German standards and EN standards; it contains measuring principles for photovoltaic, system components, tests, PV-power converters and grid connection.

- BS 7671 Requirements for electrical installations, IEE Wiring Regulations
- BS 6626 Code of practice for maintenance of electrical switchgear and control gear for voltages above 1 kV and up to and including 36 kV
- BS EN 60947 Specification for low voltage switchgear and control gear

- BS EN 60529 Specification for degrees of protection provided by enclosures (IP code)
- BS EN 62305 Code of practice for protection of structures against lightning.

German standards DIN EN

These standards are largely based on IEC/TS standards, with the exception of the following standards:

- DIN EN 60891 Procedures for temperature and irradiance corrections to measured I–V characteristics of crystalline silicon photovoltaic devices
- DIN EN 50548 Junction boxes for photovoltaic modules
- DIN CLC/TS 50539 Low-voltage surge protective devices – Surge protective devices for specific application including d.c.
    - *Part 12* Selection and application principles – SPDs connected to photovoltaic installations
- DE-AR-E-2283-4 Requirements for cables for PV systems
- DIN EN 62305 Protection against lightning
    - *Part 3* Physical damage to structures and life hazard – Supplement 5: Lightning and overvoltage protection for photovoltaic power supply systems.

Australian Standards and New Zealand Standards (AS/NZS)

The AS/NZS standards related to solar farm PV system design and installations include:

- AS/NZS 5033 Installation and safety requirements of PV arrays
- AS/NZS 3000 Electrical wiring rules
- AS/NZS 3008 Electrical installations – Selection of cables
- AS/NZS 4777 Grid connections of energy systems via inverters.
- AS/NZS 1768 Lightning protection
- AS/NZS 1170.2 Wind loads
- AS/NZS 1664.1 Aluminium structures
- AS/NZS 4600 Cold-formed steel structures.

Singapore

- Singapore standard 555:2010 Code of practice for protection against lightning
- Singapore standard 371:1998 Specifications for uninterruptable power supplies.

The Japanese Standard JIS

JIS are standards published by the Japanese Standards Association. Most of these standards are based on IEC/TS standards; however, some have been developed to reflect unique circumstances in Japan (Yamada, 2013).

- JIS Q 8901 Terrestrial photovoltaic (PV) modules – Requirements for reliability assurance system (Design, production and product warranty).
- JIS C 8990 series Crystalline silicon terrestrial photovoltaic (PV) modules – Design qualification and type approval (based on IEC/TS 61215)
- JIS C 8991 series Thin-film terrestrial photovoltaic (PV) modules – Design qualification and type approval
- JIS C 8992 Certification of safety conformity of PV modules
    - *Part 1:* Structure requirements
    - *Part 2*: Testing requirements.

PV design and simulation software

PVSyst

This software allows users to accurately analyse different design configurations of remote or grid-connected PV systems, to evaluate results and identify the optimal design solution, to provide the highest system performance; www.pvsyst.com/en

System Advisor Model (SAM)

SAM is a performance and financial model that makes performance predictions and cost of energy estimates for grid-connected PV projects based on installation and operating costs and system design parameters that users specify as inputs to the model; https://sam.nrel.gov

HELIOS 3D

This is a professional planning tool for utility-scale PV plants and can be used for project development, layout and engineering. It allows shadow free placement of PV racks on a digital terrain at any geographical position and at any given date or time; www.helios3d.com/index.php/en/

PV*SOL premium

This software enables users to plan and visualise ground-mounted systems up to 2 MW. It also allows for detailed shading analysis for yield analysis; www.valentin-software.com/en/products/photovoltaics/57/pvsol-premium

PVWatts

This tool estimates the electricity production of a grid-connected roof- or ground-mounted photovoltaic system based on a few simple inputs; http://pvwatts.nrel.gov

HELIOSCOPE

This software uses solar specific CAD-like tools and advanced energy simulation tools for the design and modelling of grid-connected PV systems up to 5 MW; https://helioscope.folsomlabs.com

PVMapper (US)

This is a utility-scale solar site mapping web application that is open-source and free to use; http://pvmapper.org/pvmapper-description.html

PVMapper Site Designer (US)

This is a free desktop extension of PVMapper that can be used to design and analyse a specific site and evaluate the amount of power that can be produced from that site; http://pvmapper.org/site-description.html

Solergo

This software aids in the complete design of grid-connected and stand-alone photovoltaic systems, developing the required documentation; www.electrographics.it/en/products/solergo.php

Ampère Professional

Electrical network calculator used to analyse the LV or MV electrical networks in AC or DC of large-scale PV systems (cables, SPD, protection, inverter) according to IEC/TS 60364, IEC/TS 60909, IEC/TS 62271-100 and international standard NF-C (France), UNE (España), BSI (UK), NBR (Brazil); www.electrographics.it/en/products/ampere_professional.php

SolarPro

This software can be used to simulate electricity generation of a PV system under different conditions and allows for system design using precise data. It can be used for systems using up to 400 inverters and 160,000 PV modules; www.lapsys.co.jp/english/products/pro.html#Specifications

Skelion

This is a plugin for Sketchup that allows users to insert 3D models of the solar installation and export energy reports automatically from PVWatts (in the USA) or PVGIS (in the EU); http://skelion.com/

PVGIS

Free online tool to estimate the solar electricity production of stand-alone or grid-connected PV systems and plants in Europe, Africa and Asia. PVGIS also provides a large and accurate solar radiation database and maps; http://photovoltaic-software.com/pvgis.php

AREMI

Australia specific data aggregation tool for map-based access to special data that can be used for planning and siting utility-scale PV plants; http://nationalmap.gov.au/renewables/

DigSilent Powerfactory

This is a tool that can used for network studies, covering generation, transmission, distribution and industrial systems; www.digsilent.com.au/

Industry magazines and periodicals

Sun and Wind Magazine

This magazine includes good general information on current developments in renewable energy; www.sunwindenergy.com

PVTECH

An industry-sponsored website featuring interesting articles on the latest developments in PV technology; www.pv-tech.org

Semiconductor Today

This website features news on the latest developments in the electronics industry which includes photovoltaics; www.semiconductor-today.com

SolarPro

This magazine publishes good information on solar technology including design and installation, products and equipment, operations and maintenance, business and finance, and project profiles. It is tailored towards professionals working in the North American solar market but much of the information is applicable worldwide; http://solarprofessional.com

Greentech Media

This website features news and discussion on current and emerging 'green' technologies including photovoltaics; www.greentechmedia.com

Renewable Energy World

A magazine focused on a wide range of topics relevant to renewable energy, including company and product information, job advertisements, blogs, podcasts and news on global developments within the industry. Both print and PDF versions are available; www.renewableennergyworld.com

PV Magazine

A monthly trade publication for the international PV community, it has independent, technology-focused reporting, concentrating on the latest PV news, topical technological trends and worldwide market developments; www.pv-magazine.com

Other PV resources

International

- International Solar Energy Society; www.ises.org
- The Renewable Energy and Energy Efficiency Partnership; www.reeep.org
- The International Renewable Energy Agency; www.irena.org/
- International Energy Agency: www.iea.org

Australia

- Australian PV Institute; www.apvi.org.au
- Australian Solar Energy Society; www.auses.org.au
- Clean Energy Council (Australia); www.cleanenergycouncil.org.au
- Department of Climate Change and Energy Efficiency; www.climatechange.gov.au
- Global Sustainable Energy Solutions (GSES), training provider; www.gses.com.au

Canada

- Canadian Solar Industries Association; www.cansia.ca

Republic of Ireland

- Department of Communications, Energy and Natural Resources; www.dcenr.gov.ie/Energy/Sustainable+and+Renewable+Energy+Division/

South Africa

- National Energy Regulator of Southern Africa; www.nersa.org.za
- Sustainable Energy Society of Southern Africa; www.sessa.org.za

UK

- Centre for Alternative Technology (Wales), training provider; www.cat.org.uk
- Department of Energy and Climate Change; www.decc.gov.uk
- Green Dragon Energy, training provider; www.greendragonenergy.co.uk
- UK Solar Energy Society; www.uk-ises.org

US

- American Solar Energy Society; www.ases.org
- Contractors License Reference Site; contractors-license.org
- Council of American Building Officials and Code Administrators; www.bocai.org
- Database of State Incentives for Renewable Energy and Energy Efficiency, information on federal, state, local and utility incentives for renewable energy; www.dsireusa.org
- Florida Solar Energy Council; www.fsec.ucf.edu/en
- Illinois Solar Association; www.illinoissolar.org
- Interstate Renewable Energy Council (IREC); www.irecusa.org
- Midwest Renewable Energy Association, training provider; www.the-mrea.org
- North American Board of Certified Energy Practitioners (NABCEP); www.nabcep.org
- Oregon Solar Association; www.oseia.org
- Solar Energy International, training provider; www.solarenergy.org
- Solar Living Institute, training provider; www.solarliving.org

Glossary

Air mass – The amount of atmosphere that solar radiation must pass through before reaching the surface of the earth.

Albedo – The scattering of reflected light back into space by gases in the atmosphere.

Alternating current (**AC**) – Electricity in which the polarity of the current is periodically reversed.

Altitude – The height of the sun from the horizon.

Ambient temperature – The temperature of the surrounding environment.

Amorphous (**silicon**) – A non-crystalline substance. Amorphous solids lack order and structure in their molecular composition. Glass is an example of an amorphous solid; so is amorphous silicon, which is used in some thin-film PV cells.

Ampere (**amp or A**) – The unit of measurement for electrical current.

Azimuth – The east–west position of the sun. The solar industry standard is to express azimuth clockwise from true north (0°–360°); however, it can also be quoted with a direction, east or west (i.e. 0°–180°E or 0°–180°W).

Back-feed current – The inverter back-feed current is the amount of current that the inverter could feed into the DC array cable during a fault condition.

Balance-of-system (**BoS**) **equipment** – The components of a PV system excluding the PV modules and inverter. BoS equipment includes the mounting structure, cabling, disconnects/isolators, module junction boxes, PV combiner boxes, grounding/earthing equipment and meters.

Battery – A container consisting of one or more cells, in which chemical energy is converted into electricity and used as a power source.

Capacity factor – The ratio of a generator's actual output over a period of time, to its potential output if it were to operate at full nameplate capacity continuously during the same period of time.

Cardinal directions – The four cardinal directions (or cardinal points) are the directions of north, east, south and west.

CCTV (**closed-circuit television**) – A TV system in which signals are monitored, primarily for surveillance and security purposes.

Cell efficiency – The amount of electrical power produced by the cell per amount of light energy hitting the cell.

Central inverter – Usually a collection of small inverters centrally located in one housing.

Circuit – A circuit is the path that current flows from one charged point to another.

Circuit breaker – A mechanical device which will open a circuit under fault conditions. When too much current passes through, the device will open and prevent current flow. The circuit breaker can then be manually operated to close the circuit.

Combiner box – An electrical component for combining and housing the wiring from the PV array.

Condition-based – Includes maintenance activities that are scheduled based on the results from system monitoring.

Corrective maintenance – Includes maintenance activities that are performed in the event of failures within the system.

Current – Current is the movement of charge and is measured in amperes (A). Conventional current is the 'flow' from positive to negative, and is opposite to electron flow. It is used throughout this publication.

Curtailment – The situation when a generator must restrict energy output because the grid can't handle all the generation (i.e. not enough demand).

DC/DC converter – A device which is connected to each solar module to ensure that each module operates at its maximum MPP (also known as a solar/ power optimiser or module MPPTs).

Debt finance – A mechanism for raising capital which comes in the form of loans, and needs to be repaid with interest throughout the project's lifetime.

Degradation – The reduction in performance of a PV module over a period of time. Usually quoted in terms of annual degradation.

Developer – The organisation that defines the scope and design of the project, and usually contracts for its planning, construction and commissioning.

Diffuse radiation – Scattered radiation that still reaches the earth's surface.

Direct current (DC) – Electricity in which the current always moves in the same direction.

Direct radiation – Radiation that passes straight through the atmosphere to the earth's surface.

Dispatchable generator – A generator that can supply electricity on demand, as opposed to those such as solar which rely on an energy source (sunlight) which is not continuously available.

Distribution transformer – In a utility-scale PV system, the distribution transformer refers to the transformer connected after the inverters to step the voltage up to medium-voltage levels for transmission to the substation. Capacity can vary between 50 and 2,500 kVA.

Due diligence – The process of evaluating the project prior to finalisation of finance agreements. It usually includes an evaluation in terms of the legal aspects, permits, contracts and specific technical issues, in order to determine the risks and mitigation methods associated with the project.

Electromagnetic radiation (EMR) – Energy that travels through space as a wave. Sunlight is an example of EMR.

Energy – The amount of electric energy transferred, a product of power and time. Energy is measured in watt hours (Wh) and is calculated using: $E = P \times t$

Energy yield – A photovoltaic system's energy output measured over a given period of time is its yield. This value is a common measurement as it is used to indicate performance as well as calculating revenue.

Engineering, procurement and construction (EPC) contract – A contract between the developer and a contractor who is responsible for the project design and engineering, the procurement of power generation equipment, construction management and possibly O&M.

Equipotential bonding – Equipotential bonding (or protective earthing) involves electrically connecting earthed, conductive metalwork so that it is at the same voltage (potential) as earth throughout. This is required for safety reasons to protect people from electric shocks.

Equity finance – A method or raising capital through contributions from investors who are entitled to a share of the income generated from the project.

Extra low voltage (ELV) – Electrical systems that operate under 120 V DC (ripple free) or 50 V AC. ELV systems do not require an electrical license to be installed or worked on (*see* **low voltage**).

Feed-in tariff (FiT) – An incentive scheme that provides fixed revenue to solar farm developers per MWh of electricity generated.

Fill factor – Fill factor partially describes the performance of the cell. It indicates how close the maximum power voltage and current are to the open circuit voltage and short circuit current.

Fixed tilt – Used to describe a module or array that is set at a fixed angle and does not change.

Frequency – An indication of the imbalance between supply and demand on the AC grid.

Functional earthing – A type of earthing designed to ensure optimal performance of the PV array, but it is required only if specified by the manufacturer.

Fuse – A device that protects conductors from excessive current. The fuse is rated to carry a certain current, and when this current is exceeded the fuse will open the circuit (by melting).

Future value – The value of an amount of money in the future, which is dependent on the length of time until the future date, and interest rate.

Global radiation – The total solar radiation that reaches the earth's surface, made up of both direct and diffuse radiation.

Grid parity – Grid parity is the point where an alternative energy source – in this case large-scale solar – can generate power at a levelised cost of electricity (LCOE) less than or equal to purchasing conventional electricity from the grid, without the need for subsidies.

Grid-connected PV system – A PV system which exports electricity to an electrical grid.

Ground fault detection interrupter (GFDI) – A safety device consisting of an amp breaker used for PV applications. When a ground fault is detected, the amp breaker will trigger all other breakers connected to GFDI to trip and shut off the power.

Ground-mounted – Describes a solar array which is supported by the ground, rather than by building roofs.

Inflation – The rate at which the value of money decreases over time.

Insolation – The amount of solar radiation incident on a surface over a day, measured in peak sun hours (PSH) or kWh/m^2/day

Intermittency – *See* **variability.**

Inverter – A device that converts DC electrical power into AC, the inverter is an essential component in grid-connected photovoltaic systems.

Irradiance – The total amount of solar radiation available at any point in time per unit area and is measured in W/m^2 or kW/m^2. It is a measure of power.

Irradiation – The total amount of solar radiation available per unit area over a specified time period such as one day. It is the sum of irradiance values over a time period and is often measured in kWh/m^2/year or MJ/m^2/day. It is a measure of energy.

Islanding – The case when a distributed generator continues to supply power to a part of the electricity network (an 'island') when that part of the network is no longer supplied with utility-generated ('grid') power.

I–V curve – A graph used to plot the output characteristics of a PV cell. The plot shows voltage versus current and can be used to determine power output and efficiency.

Junction box – A box containing a junction of electric wires or cables.

Levelised cost of electricity (LCOE) – A common metric used to quote and compare the cost of electricity produced by a generator, expressed in terms of $/MWh.

Low voltage (LV) – Electrical systems that operate between 50 and 1,000V AC or 120 and 1,500V ripple-free DC.

Magnetic declination (deviation) – The difference between true north (the direction of the north pole) and magnetic north (the direction which a compass will point).

Maximum power point (MPP) – The point on the I–V curve that gives the maximum power. It occurs when the load resistance is equal to the internal resistance of the PV cell.

Maximum power point tracker (MPPT) – An electronic device included within the inverter that alters the PV array's electrical output so that it is performing at the maximum power possible at any given time.

Micro-inverter – An inverter that is designed to be mounted on the back of a module.

Module efficiency – The amount of electrical power produced by the module per amount of light energy hitting the module. This is typically lower than cell efficiency due to losses from reflection from glass, etc.

Monocrystalline solar cells – The most efficient and most expensive solar cells. They have a smooth monochromatic appearance.

Multi-crystalline solar cells (polycrystalline) – Less efficient solar cells, but cheaper to make and buy. They have a 'glittering' effect when in sunlight.

Multi-string inverter – An inverter with multiple MPPT's (e.g. one MPPT per string).

Nominal – A nominal value is a reference value used to describe batteries, modules or systems. Therefore, it is not an exact value. For example, the nominal voltage of a 72-cell solar module is 24 V but the open circuit voltage (V_{OC}) for the same module can be 45.6 V and the maximum power voltage (V_{MP}) can be 35 V.

Open circuit – An open circuit is where the current path is broken so that the current is equal to 0.

Peak power – The maximum amount of power a PV cell, module or array is expected to produce under standard test conditions. Peak power is typically given in watts-peak (Wp), kilowatts-peak (kWp) or megawatts-peak (MWp).

Peak sun hours (PSH) – A unit of energy used in the solar industry when measuring irradiation; 1 PSH = 1 kW of solar energy falling on a surface of 1 m^2 for 1 hour.

Performance ratio – The ratio between the actual energy yield from a solar array and maximum theoretical output of the array.

Photovoltaic (PV) – A device which creates electricity when sunlight hits its surface.

Polycrystalline solar cells – *See* **multi-crystalline cells.**

Power – The rate at which electric energy is transferred. Power is measured in watts (W), and is calculated using $P = V \times I$.

Power curtailment – In the event that there is too much supply to meet the local electricity demand, the generator may have to reduce (curtail) power output in order to maintain grid stability.

Power off-taker – The buyer (usually an electricity utility) of electricity generated from a power plant sold under a power purchase agreement (PPA).

Power purchase agreement (PPA) – An agreement between the plant owner and the power off-taker for the sale of electricity generated by the plant for a fixed price over a specified period of time.

Present value – The value of an amount of money now, which would equate to a desired amount of money in the future dependent upon length of time and interest rate.

Preventative maintenance – Includes scheduled maintenance activities carried out to prevent issues occurring with the system equipment that may lead to reduction in performance and loss of revenue.

PV array – Strings of PV modules are electrically connected in parallel to form an array (also called a solar array).

PV cell – A single PV device (also called a solar cell).

PV module – PV cells are physically and electrically connected to form a PV module. These cells are held together by a frame and covered by a protective substance such as glass (also called a solar module).

PV string – The term used to describe a group of PV modules connected in series.

PV sub-array – Very large PV arrays are often made up of many smaller PV arrays known as sub-arrays.

PV system – The PV array and all associated equipment required to make it work (also called a solar electric system).

Pyranometer – A device used to measure global horizontal solar radiation (W/m^2) from a 180° field of view angle.

Ramp rate – The rate at which a generator changes its output expressed in MW/min.

Reactive power – Reactive power exists when voltage and current are not in phase. It is expressed in terms of volt-ampere reactive (VAR).

Residual current device (RCD) – A current-activated circuit breaker used as a safety device for mains-operated electrical tools and appliances.

Resistance – The opposition to current and is measured in ohms (Ω).

Root mean square – Is how AC power is usually quoted; for example: $V_{RMS} = 0.707 \times VP$ … $I_{RMS} = 0.707 \times IP$

Short circuit – Where the current is flowing in a closed path across the source terminals.

Short-run marginal cost (**SRMC**) – An economic term referring to the operating costs of the power plant.

Soiling – The presence of dirt, dust, bird droppings etc. on solar modules and reduces the efficiency of the array.

Solar altitude (**elevation**) **angle** – The angle between the sun and the horizon. This angle is always between 0° and 90°.

Solar azimuth angle – The angle between north and the point on the compass where the sun is positioned on a horizontal plane. The azimuth angle varies as the sun moves from east to west across the sky throughout the day. In general, the azimuth is measured clockwise going from 0° (true north) to 359°.

Solar cell – A small photovoltaic unit that generates an electrical current when hit by sunlight.

Solar modules – *See* **PV module.**

Solar noon – The time of day when the Sun is exactly half way between sunrise and sunset: when the Sun is at its highest point for that day.

Solar radiation – Energy coming from the Sun.

Spinning reserve – Back up generating capacity that is available to the transmission system at short notice by increasing the power output of generators.

Stand-alone PV system – A PV system which provides or supplements the main electrical supply. These systems use batteries to store power.

Standard test conditions (**STC**) – Standardised test conditions which make it possible to conduct uniform comparisons of PV modules by different manufacturers.

String inverter – An inverter with only one MPPT.

Substation – The area of a solar farm that houses the substation transformer, switchgear, protection systems, monitoring and control systems, metering equipment and other associated equipment.

Substation transformer – In a utility-scale PV system, the substation transformer refers to the transformer located at the substation, used to step the voltage up to high-voltage levels for connection to the utility grid. Capacity can vary from 2,500 kVA to more than 100 MVA .

Supervisory control and data acquisition (**SCADA**) **systems** – The system used to monitor and control various components of the PV system.

Surge protection – Devices or appliances that are used to protect electrical devices from voltage spikes.

Thin-film solar cells – Solar cells made from materials that are suitable for deposition on large surfaces such as glass. Very thin in comparison to monocrystalline and multi-crystalline solar cells. Least efficient technology, but the cheapest to manufacture.

Tracking systems – Mounting systems that include mechanical devices which can alter the orientation axis and/or tilt axis of the PV array in order to optimise the exposure of the array to the sun and capture more solar radiation.

Transformer – An electrical device used to step voltage up or down between at least two electrical circuits.

Transient current – An oscillatory or aperiodic current that flows in a circuit for a short time following an electromagnetic disturbance such as a nearby lightning strike.

Uninterruptible power supply – Provides short-term power backup for important applications, such as IT and telecommunications.

Variability – The non-continuous nature of some generators that depend on fuel sources that vary with climatic factors, such as sunlight.

Voltage – The potential difference between two points; measured in volts (V).

Watt peak (Wp) – A non-SI unit used in the solar industry to describe the nominal power of a solar PV system: it refers to the peak output under standard test conditions. Kilowatt-peak (kWp) and megawatt-peak (MWp) are commonly used to describe the output of a PV array.

Index

Page numbers in **bold** denote tables and in *italics* denote figures.

Milton Keynes UK
Ingram Content Group UK Ltd.
UKHW050452071024
449327UK00015B/340